T0225704

Advanced Technologies and Societal Change

Series Editors

Amit Kumar, Bioaxis DNA Research Centre (P) Ltd., Hyderabad, Telangana, India

Ponnuthurai Nagaratnam Suganthan, School of Electrical and Electronic Engineering, Nanyang Technological University, Singapore, Singapore

Jan Haase, NORDAKADEMIE Hochschule der Wirtschaft, Elmshorn, Germany

Editorial Board

Sabrina Senatore, Department of Computer and Electrical Engineering and Applied Mathematics, University of Salerno, Fisciano, Italy

Xiao-Zhi Gao⑩, School of Computing, University of Eastern Finland, Kuopio, Finland

Stefan Mozar, Glenwood, NSW, Australia

Pradeep Kumar Srivastava, Central Drug Research Institute, Lucknow, India

This series covers monographs, both authored and edited, conference proceedings and novel engineering literature related to technology enabled solutions in the area of Humanitarian and Philanthropic empowerment. The series includes sustainable humanitarian research outcomes, engineering innovations, material related to sustainable and lasting impact on health related challenges, technology enabled solutions to fight disasters, improve quality of life and underserved community solutions broadly. Impactful solutions fit to be scaled, research socially fit to be adopted and focused communities with rehabilitation related technological outcomes get a place in this series. The series also publishes proceedings from reputed engineering and technology conferences related to solar, water, electricity, green energy, social technological implications and agricultural solutions apart from humanitarian technology and human centric community based solutions.

Major areas of submission/contribution into this series include, but not limited to: Humanitarian solutions enabled by green technologies, medical technology, photonics technology, artificial intelligence and machine learning approaches, IOT based solutions, smart manufacturing solutions, smart industrial electronics, smart hospitals, robotics enabled engineering solutions, spectroscopy based solutions and sensor technology, smart villages, smart agriculture, any other technology fulfilling Humanitarian cause and low cost solutions to improve quality of life.

Kandiah Pakeerathan
Editor

Smart Agriculture for Developing Nations

Status, Perspectives and Challenges

 Springer

Editor
Kandiah Pakeerathan
Department of Agricultural Biology
University of Jaffna
Jaffna, Sri Lanka

ISSN 2191-6853 ISSN 2191-6861 (electronic)
Advanced Technologies and Societal Change
ISBN 978-981-19-8740-3 ISBN 978-981-19-8738-0 (eBook)
https://doi.org/10.1007/978-981-19-8738-0

© Centre for Science and Technology of the Non-aligned and Other Developing Countries 2023
This work is subject to copyright. All rights are solely and exclusively licensed by the Publisher, whether the whole or part of the material is concerned, specifically the rights of translation, reprinting, reuse of illustrations, recitation, broadcasting, reproduction on microfilms or in any other physical way, and transmission or information storage and retrieval, electronic adaptation, computer software, or by similar or dissimilar methodology now known or hereafter developed.
The use of general descriptive names, registered names, trademarks, service marks, etc. in this publication does not imply, even in the absence of a specific statement, that such names are exempt from the relevant protective laws and regulations and therefore free for general use.
The publisher, the authors, and the editors are safe to assume that the advice and information in this book are believed to be true and accurate at the date of publication. Neither the publisher nor the authors or the editors give a warranty, expressed or implied, with respect to the material contained herein or for any errors or omissions that may have been made. The publisher remains neutral with regard to jurisdictional claims in published maps and institutional affiliations.

This Springer imprint is published by the registered company Springer Nature Singapore Pte Ltd.
The registered company address is: 152 Beach Road, #21-01/04 Gateway East, Singapore 189721, Singapore

Foreword by Leela Devi Dookun-Luchoomun

It was an honour for the Republic of Mauritius, through my Ministry, to host in a virtual mode and in collaboration with the Centre for the Science and Technology of the Non-Aligned and Other Developing Countries (NAM S&T Centre), the International Workshop on *Smart Agriculture for Developing Nations: Broader Perspectives and Special Challenges for the Island States.*

The subject could not have been more opportune. According to the Food and Agricultural Organization, the agricultural sector is very likely to face enormous challenges in feeding the 9.6 billion prospective inhabitants of the planet by 2050. Food production must increase by 70% by 2050, despite limited arable land, coupled with the decreasing availability of fresh water. The situation can be potentially more drastic for small island developing states whose vulnerability to prolonged periods of droughts and floods caused by climate change increases by the day. As a small island developing state, Mauritius faces several challenges in ensuring access to nutritious food at affordable costs to its population while eliminating food wastage and minimizing post-harvest losses.

It thus becomes imperative for SIDS to upgrade their traditional farming practices. There is absolutely no doubt that climate-smart, data-driven and data-enabled agriculture is the way to go: they must have recourse to a tech revolution to develop and adopt eco-friendly systems and climate-smart techniques for enhancing crop productivity while, simultaneously, encouraging the emergence of young agricultural entrepreneurs.

It is my firm conviction that extensive research would revolutionize the agro industry and lead to the adoption of innovative techniques in crop production. We have already witnessed precision agriculture's benefits of minimizing water, herbicides, pesticides, and fertilizers. Similarly, smart agriculture allows the crops to be monitored constantly and helps detect an early sign of a disease outbreak. It further helps in sharing information crucial for farmers and enhances access to technical details.

This virtual workshop has certainly helped the NAM S&T Centre to fully meet its objectives of rallying a number of stakeholders and experts in the field for them to strengthen their collaborative network and pool their research-based knowledge and share ideas and practices. In many cases, these are food for thought for more intense and impactful research.

After all, lessons are there to be learnt and new techniques implemented primarily for humanity's benefit.

I am sure readers will enjoy the knowledge contained in this book following this vital workshop on smart agriculture.

I wish you all a pleasant read.

Hon. Mrs. Leela Devi
Dookun-Luchoomun
Vice Prime Minister, Minister
of Education,
Tertiary Education, Science
and Technology
Vacoas-Phoenix, Republic of Mauritius

Preface

Anthropogenic activities and rapid industrialization around the globe have caused catastrophic climate change which poses constant threat to global food security and sustainable development. Due to the impact of ever-increasing human population, greenhouse gas emissions, and temperature, many countries started to experience unpredicted heavy flooding, erratic rainfall associated with long dry spells and/or drought, changes in average temperatures, heat wave strikes, frost, cyclones, hailstorms, and dust storms that not only ruin the agriculture but also lower crop productivity and yield. Apart from the climate, traditional methods of intensive agriculture have over exploited the natural resources, depleted the soil fertility, diminished the fresh water resources, increased the land fragmentation, disturbed the natural balance, and increased pollution. These calamities have also led to poor crop establishment and increased pest and disease attack. In many developing countries, food security and rural development are facing many challenges such as migration of rural population to cities resulting in labour shortage for agricultural production. Shortage of labour at times of peak demand is another limiting factor to expand extent of cultivation in developing countries especially the small island nations. Post-harvest processing and marketing are also limited in the developing countries to make use of technologies needed for value addition.

Considering the above challenges, transformation of conventional agriculture towards "Smart Agriculture" is the only option to improve the farming practices for efficient use of available resources, eliminate the risk of losing yields, optimize the control over production, better cost management, waste reduction and eliminate hunger. Smart farming helps farmers to better understand the important factors such as water, topography, vegetation and soil types and provides options to utilize advanced technologies like artificial intelligence (AI), big data, robotics, drones, and Internet of Things (IoT) to increase crop yield with efficient utilization of precise natural resources without harming the environment. This allows farmers to determine the best uses of scarce resources within their production environment and manage these in an environmentally and economically sustainable manner.

In the post "COVID-19 pandemic" era, multiple economic crises situation and prevalence of food shortage and insecurity highlight the importance of raising crop yields and adaptation to climate change. The publication of this book is timely as it is essential to create interest and awareness of private and public sectors, policy-makers and investors on the importance of transformation of the developing nations' traditional agriculture towards the "Smart Option". The current smart farming techniques shared in different chapters will not only help to reduce the overall cost and improve the quality and quantity of products but will altogether help improve the sustainability of agriculture in the developing world.

The present book titled *Smart Agriculture for Developing Nations: Status, Perspectives and Challenges* being published by Springer Nature, Singapore, is an initiative of the Centre for Science and Technology of the Non-Aligned and Other Developing Countries (NAM S&T Centre), New Delhi. The book has a total of 18 chapters contributed by the authors from India, Indonesia, Iraq, Malaysia, Mauritius, Myanmar, Palestine, Sri Lanka, and South Africa that intend to provide sustainable and smart technology-based solutions for developing nations. I am sure that the book will be a useful resource guide for research scientists, policy-makers, and students to understand the present scenario of "Smart Agriculture in Developing Countries and Island Nations" and plan future researches to minimize the knowledge and technology gaps present today in the area especially amongst the scientific community in the NAM and other developing countries.

Dr. Kandiah Pakeerathan
Department of Agricultural Biology
University of Jaffna
Jaffna, Sri Lanka

Introduction by Amitava Bandopadhyay

According to the UN estimates, by 2050 there will be 9.7 billion people in the world—in other words around 2 billion more mouths to feed than in 2020. This increase in population needs to be met through a 70% rise in agricultural production.

The situation poses a serious challenge to the developing countries with regard to the "Sustainable Development Agenda—2030" and specifically SDG-2, which aims *to end hunger and provide access to safe, nutritious and sufficient food to all while ensuring sustainable food production systems and implementing resilient agricultural practices that increase efficiency.*

In addition, the food industry is currently responsible for 30% of the world's energy consumption and 22% of greenhouse gas emissions. The challenge, therefore, is not just producing more food, but doing it in a sustainable manner.

The concept of smart farming has arisen at the dawn of the "Fourth Industrial Revolution" in the areas of agriculture to increase the production quantity as well as quality, by making maximum use of resources while minimizing the environmental impact.

"Smart farming" is a concept that uses technologies like drones to monitor hundreds of acres of land to assess the health of crops and cattle, smart sensors to help in the early detection of infestations and automatic systems to water, fertilize and fumigate each plot depending on its specific characteristics and on the basis of weather forecast. These are just some of the virtues of smart farming which may help eradicating hunger in an over-populated future.

Increased investment through enhanced international cooperation, agricultural R&D and smart technological innovations to enhance agricultural productivity and capacity is significant for developing countries, particularly for the Least Developed Countries (LDCs).

Also, in the light of the effects of COVID-19 pandemic on the food and agricultural sector, prompt measures are needed to ensure that food supply chains are kept alive to mitigate the risk of large shocks that have a considerable impact on the society.

This book through its 18 chapters intends to provide smart and sustainable technology-based agricultural solutions to the developing countries. It attempts to ensure that the implementation of technology in agriculture is significant for developing countries to boost their food security and safety while ensuring the overall sustainability of agriculture sector. Some of the chapters specifically deal with special challenges in agriculture related to the Small Island Developing States (SIDS).

The book brings together scientific communities from India, Indonesia, Iraq, Malaysia, Mauritius, Myanmar, Palestine, Sri Lanka and South Africa to share their expertise to make the farming system more connected and intelligent. The smart farming techniques shared in different chapters will not only help to reduce the overall cost and improve the quality and quantity of food production, but will also help to maintain the food supply chain sustainability.

I am honoured by the support extended by Her Excellency Mrs. Leela Devi Dookun-Luchoomun, Vice Prime Minister, Minister of Education, Tertiary Education, Science and Technology, Republic of Mauritius, and writing the "Foreword" of this book.

I am thankful to Dr. Loyola D'Silva, Executive Editor, Springer Nature, Singapore for his kind support in bringing out this publication and Ms. Saranya Devi Balasubramanian, Project Coordinator, Springer Nature, India, for managing all technical and administrative tasks of the publication process.

I am grateful to Dr. Kandiah Pakeerathan, Head of Department of Agricultural Biology, Faculty of Agriculture, University of Jaffna, Kilinochchi, Sri Lanka, for taking the responsibility to serve as Editor of this book. Dr. Pakeerathan's role has been essential in complementing the scope of this book, and his expertise and guidance have been instrumental in the selection of the best articles that are published in this book.

I also acknowledge the valuable support of the entire team of the NAM S&T Centre and I am especially thankful to Mr. Madhusudan Bandyopadhyay, Senior Adviser; Ms. Jasmeet Kaur Baweja, Programme Officer; Ms. Rhaeva Bhargava, Research Associate, and Ms. Abhirami Ramdas, Research Assistant, for their contributions in taking this publication project forward and bringing it to a successful conclusion.

I also record my appreciation for the assistance and support rendered by my colleagues Mr. Rahul Kumra and Mr. Pankaj Buttan towards bringing out this publication.

I hope that this publication would be a useful reference material for scientists and researchers from R&D institutions, technology providers and start-ups, agribusiness professionals, farming/agriculture technology consultants, project planners, government officials, policy-makers, and other representatives from industry and non-government organizations who are actively engaged in areas of smart agriculture.

Amitava Bandopadhyay, Ph.D.
Director General
NAM S&T Centre
New Delhi, India

Contents

Chapter 1
Smart Agricultural Mechanization in India—Status and Way Forward

C. R. Mehta, N. S. Chandel, and Kumkum Dubey

Introduction

In the current digital era, Information Communication Technology (ICT) is proliferating with smart/intelligent machines which can further advance the technologies in industrial and agriculture sectors. Issues concerning agriculture have been always hindering the development of the country. The only solution to this problem is smart agriculture which streamlines the traditional methods of agriculture. In order to modernize the agri-technology and precision farming (PF), there is a need of new hi-tech computing approaches like Internet of Things (IoT), Artificial intelligence (AI), big data, robotic system, unmanned ground vehicle (UGV), unmanned aerial vehicle (UAV), etc. [10]. This digital farming will lay the foundation for the "third green revolution" in India. With the introduction of smart faming technology in agriculture sector, the impact of weather conditions, use of fertilizers and water reduces, however, worker safety, efficiency, and production rate increases. It also helps growers to manage and control production threats and improve the capability to foresee process outcomes. IoT-based smart farming is a system that is built for monitoring the crop field with the help of sensors (light, humidity, temperature, soil moisture, crop health, etc.) and automating the field operations [1, 7]. IoT connects sensors, actuators, and controller with server by accessing Internet connection (Fig. 1.1). Thus, farmers can monitor the field conditions from anywhere with the help of IoT technology. They can also select between manual and automated options for taking necessary actions based on sensor data. AI-based analytical or prediction models are applied on data captured from IoT devices for monitoring crops, surveying and mapping the fields,

C. R. Mehta (✉) · N. S. Chandel · K. Dubey
ICAR-Central Institute of Agricultural Engineering, Bhopal 462038, India
e-mail: cr.mehta@icar.gov.in

N. S. Chandel
e-mail: narendra.chandel@icar.gov.in

© Centre for Science and Technology of the Non-aligned and Other
Developing Countries 2023
K. Pakeerathan (ed.), *Smart Agriculture for Developing Nations*, Advanced Technologies
and Societal Change, https://doi.org/10.1007/978-981-19-8738-0_1

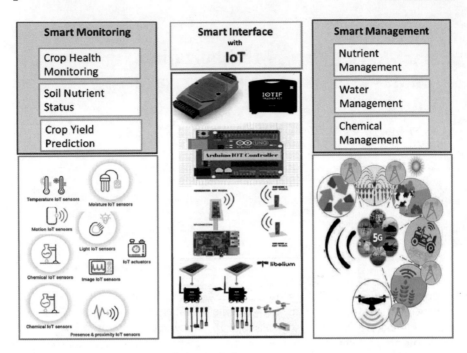

Fig. 1.1 Architecture of IoT-based smart agriculture

identify abiotic/biotic stress, predicting environmental conditions, and crop yield [4]. Such type of smart farming technologies save both time and money and also alert the farmers to take necessary action before crop damage. With the hybridization of AI and IoT approaches, autonomous robotic machines can be developed which solve major field issues like crop diseases infestations, pesticide and fertilizer control, grain storage management, irrigation control, and complex crop harvesting practices [12, 14].

Smart farming emphasizes the concept of digital agriculture using AI, IoT, remote sensing, robotics, cloud computing, and big data analytics. Such types of smart farming practices reduce total budget and improve the value, worth, and capacity of crops. It also helps to maintain the balance between supply and demand chains of agriculture products in marketplace. Thus, smart digital farming has potential to improve the agriculture sustainability and allows real-time data collection, controlling, and decision-making under several agricultural activities [8]. The decision-making can be deployed in the embedded system using sensor data as input and results are directly transferred to the machine for performing precise management actions.

Digital agriculture shifts farming's culture from 'hands-on' and experience-based management to a data-driven approach. It is an information base agriculture framework in which digital data from agricultural production and management systems are collected, processed, and interpreted. The digital agriculture revolution relies heavily on big data and cloud computing. Cloud computing stores large amounts of data at a reasonable cost and allows instant access of data. Various firms, such as AWS, Microsoft, Amazon, Google, IBM, etc., provide three types of cloud services, i.e., software, platform and infrastructure over the Internet. Cloud service is being used in different fields, however, in current era, it plays a vital role in agriculture field. Cloud computing allows immediate gathering and storing of large amount of data and also provides numerous software applications and tools for users. Farmers may use cloud-based farm management tools to store and retrieve information of market, field data, newly developed agriculture implements, weather, and climatic information [11]. This database helps framers in taking immediate managerial decisions at right time. Cloud services assist in the resolution of farmer's queries by providing immediate solutions based on domain expertise. They can seek expert advice online via cloud-based databases.

Farm management system is a cloud-based program that provides online training for farmers to educate them about diseases and pesticides, as well as how to maintain a farm [3]. It also gives information regarding new software application of farming practices or machinery. IoT uses cloud services for storing and accessing of sensors/devices data installed in field. The cloud service also helps farmers in preventing crop damage due to climate change, as farmers receive weather reports and take appropriate action to save the crop. Big data use cloud database for predicting/forecasting different conditions of farm activities as well as environment and automate the machines according to outcomes generated from analysis tools. Big data can truly revolutionize the agricultural sector only by having a cloud-based ecosystem with the right tools and software to integrate various data sources. The opportunity for applications of big data in the present agriculture is large. The capability to track physical items, gather real-time data, and forecasting events can be a real game changer in digital farming. The big data analytics contribute and really transform the agricultural sector in feeding a growing population, using pesticides ethically, optimizing farm equipment performances, managing supply chain issues and much more to aid the farmer take decisions.

Technologies for Smart Agriculture

Artificial Intelligence (AI) in Agriculture

Artificial intelligence (AI) enables the machine to simulate like a human intelligence and provide intricate details of the problem, learns from external data, and uses those learning to provide a best-fit solution through flexible adaptation [10]. Nowadays, very complex linear/non-linear problems are being solved with this technique. AI is becoming pervasive very rapidly in agriculture sector because of its robust applicability in the complex problems of farming that cannot be solved well by humans and traditional farming approach. Different applications of AI in agriculture sector are described below.

Fig. 1.2 Device for the
identification of soybean
crop disease

Detection of Abiotic and Biotic Crop Stresses Using AI Models

The timely detection of water stress in crop is one of the major challenges in precision and sustainable agriculture. Hence, abiotic stress detector module has been developed at ICAR-Central Institute of Agricultural Engineering (CIAE), Bhopal, to identify water or nitrogen stress in crop. The AI-based system uses deep learning model to detect or to classify stress/non-stress crop in real-time [5]. This module helps in detecting the abiotic stress at early stages, so that farmers can take necessary actions for controlling the stress and saving the crop from further damage.

A hand-held device based on deep learning (DL) models has also been developed at ICAR-CIAE, Bhopal for identification of disease at field level in soybean crop. This device consists of a single-board microcontroller, display unit, RGB camera, DC power supply, and other accessories (Fig. 1.2). The device identifies Yellow Mosaic disease of soybean crop in less than 20 s time with 90% accuracy by using image of crop leaves. The cost of hand-held device is around Rs. 10,000/-. Deep learning model was trained by a large number of images of soybean leaves and validated, tested, and deployed in micro-controller Raspberry pi for real-time detection of crop disease. Graphical user interface was also developed in both types of modules (biotic and abiotic) so that user can easily access the software and get knowledge about the abiotic/biotic stress of crop.

Crop Monitoring and Yield Prediction

AI-based predictive models help in forecasting the crop yield by monitoring crop health at different stages of growth on real-time basis. AI models are trained by previous time series and then used for predicting the future outcome. It also helps in predicting the best time for sowing crop and also provides irrigation and fertilizer recommendation to farmers on the basis of weather conditions [4]. Such type of modern smart approach also helps in harvesting the crop timely without any damage due to adverse climatic condition. It alerts the farmers about the future adverse climate, so that farmers can timely harvest the crop.

Embedded System for Small Farms

Embedded system is the amalgamation of sensor, micro-controller, actuator, and software program. This system integrates electronics, computers, and computing technologies into bio-systems such as agriculture, forestry, and horticulture [14]. The potential benefits of embedded system-based agriculture devices are that they enhance application accuracy and operation safety in the precision farming sector. Robotic system is the further advancement of embedded system in electronics area. Robot is a machine that can perform specific task according to the program stored in its database using some mechanical claw, hand, or tool attached to the body [12]. There are two types of robotic systems, viz., remote-controlled robot and autonomous robot. The remote-controlled robots are semi-automatic, as they required human being for controlling their operation. However, autonomous robot is fully automatic; there is no need of human being for performing the task. Autonomous robot controls their action on the basis of information from sensors mounted on the machine. Various robotic systems are used in agriculture field like vision-based navigation system, harvesting robot, GPS-based vehicle guidance system, autonomous spraying unit, etc. Moreover, AI technology enhanced the efficiency of autonomous robotic system in farm practices.

A real-time uniform spraying system has been developed at ICAR-CIAE. It maintains uniform application rate of chemical throughout the field irrespective of forward speed of operation and reduces loss of chemicals during turning at headlands (Fig. 1.3). This system uses hall effect sensor for computing the speed of tractor and proportional flow control valve independent to pressure for varying the discharge rate. The spraying system has been evaluated in the field for application rate of 300 l/ha at 196 kPa pressure. With change in forward speed from 2.43 to 4.50 km/h, the application rates are in the range of 294–298 l/ha corresponding to the targeted application rate of 300 l/ha. The effective field capacity of the machine is 0.7 ha/h at forward speed of 3 km/h.

Automatic Object Detection Based Spraying System for Orchards

A tractor-operated ultrasonic sensor-based automatic pesticide sprayer was developed and tested at IIT, Kharagpur. This automatic spray control unit has been interfaced with arduino, ultrasonic sensors, pump, solenoid valves, nozzles, HTTP pump, pressure gauge, pressure relief valve, tank of 200 l capacity, and 12 V DC power supply (Fig. 1.4). The sonal signals of sensors instigated the micro-controller system for spraying in desired area. The ultrasonic sensor could detect a set object within a sensing range of 0–3 m. It efficiently sprays chemical only on orchard canopy and

Fig. 1.3 Sensor based real-time uniform rate spraying system in potato crop

Fig. 1.4 Ultrasonic sensor based pomegranate spraying system

cut-off spray when there is no plant canopy. The developed spraying system saves 26% pesticide especially in small pomegranate orchards.

Automatic Irrigation System for Rice Crop

There is a challenge of controlling irrigation water in rice field. Hence, an automatic irrigation system has been developed at ICAR-CIAE for alternate wetting and drying (AWD) method of irrigation to detect ponding water in rice fields (Fig. 1.5). It consists of sensors and micro-controller to detect water level and transmit signal to the controller wirelessly using radio frequency module. The controller unit has been programmed to operate the pump based on the desired water level in the field at different stages of crop growth.

Fig. 1.5 Automatic irrigation system for rice field

Automatic Grain Health Monitoring System

An automatic grain storage monitoring system was developed at ICAR-CIAE to monitor the micro-environment, to control insect activity and restrict spoilage of stored grain in storage system (Fig. 1.6). A sensor rod was mounted with DHT22 sensors for monitoring temperature, relative humidity, and carbon-dioxide levels at three different locations inside a storage system. In the developed embedded system, sensors, LED indicators, and LCD with a data logger are interlinked with micro-controller arduino. The developed program generates an alert message when any of the physical parameters like temperature, relative humidity, and carbon-dioxide deviate from their threshold values. The sensor rods have been tested in a flexible PVC-coated fabric bag to monitor wheat grain health for a period of 8 months. It has been observed that insect infestation activity increases the CO_2 level. It was also observed that CO_2 sensors are more suitable in detecting the grain health as compared to conventional temperature and relative humidity level of the interstitial environment of a bagged grain storage system. The detection sensitivity of CO_2 sensor used in the system is 406–5764 ppm.

Automated Packing Line for Spherical Horticultural Produces

An automated packing line for spherical horticultural produces has been developed at ICAR-CIAE for real-time sorting of produces on the basis of three weight categories and color (Fig. 1.7). Color and weight-based sorting algorithms are individually programmable to accommodate variety of spherical fruits like oranges, sweet limes, apples, etc. The packing line is attached with a water jet washer and perforated cylindrical LDPE heat sealing packing unit. The overall capacity of the machine is about 200 kg/h. The efficiencies of color and weight-based real-time sorting machine are 92 and 88%, respectively.

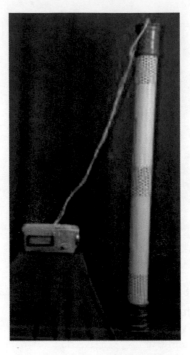

Fig. 1.6 Alert system for automatic grain health monitoring

Fig. 1.7 Automated packing line for spherical horticultural produces

Unmanned Aerial Vehicle (UAV)

UAV is the amelioration of remote sensing technology for precision agriculture (PA) and smart farming. Remote sensing is commonly used for monitoring cultivated fields and vegetation parameters at specific growth stages through images of different wavelengths. In early decades, remote sensing was often based on satellite images, however, satellite images have low spatial resolution and sometime environmental conditions hinder the image-capturing process. There is a period between acquisition and reception of satellite images. The development of UAV-based remote sensing systems has enhanced remote sensing for farm practices. The application of UAVs to monitor crops offers great possibilities to acquire field data in an easy, fast, and cost-effective way compared to traditional remote sensing methods [13]. Moreover, UAV can provide ultra-high spatial resolution of crop images at low altitude by covering a large field in a short time more efficiently than the ground systems. Equipped with sensors of different types, UAVs can identify different problems of fields and inform the farmers about their field condition, so that they can take necessary action on time for saving the crop from damage. UAV can be used in different agriculture applications such as crop health and growth monitoring, yield prediction, weed management and detection, abiotic, and biotic stresses detection. UAV provides an imagery database for training AI models or IOT-based system by mounting camera of visible or IR range on drones. UAV system can also be used for spraying in small fields on the basis of GPS and imagery data.

In current scenario, UAV technology is not only used for capturing images at different altitude but also used for spraying in small or marginal farmlands. Marut Drone Tech start-up has developed an intelligent and autonomous drone for spraying application in agriculture. The drone collects data, analyzes them and generates disease map of a particular field with the help of RGB (red, green, blue), hyper, multi-spectral cameras, and powerful sensors. Drone takes the payload with it and sprays the targeted areas using a predefined route. The start-up is presently working with input manufacturers and Farmer Producer Organizations (FPOs) to provide services [8]. It has already covered about 2000 ha of crops by targeted spraying of pesticides in Telangana and Andhra Pradesh states.

Modern Horticulture Farming Techniques

In order to enhance the innovations in agriculture sector, smart farming techniques are bringing forth new approaches that help the farmers to increase productivity and quality of the products with less use of natural resources. Nowadays, soil-less farming techniques like hydroponics and aeroponics are in vogue, as they save natural resources up to a great extent. Hydroponics means growing plants in a water and nutrient solution, without soil or any other solid media. The nutrients are provided to plants by water solution and EC, pH, and TDS of solution are maintained timely for proper growth of crop. There are various types of hydroponic systems according to their design and functioning, i.e., nutrient film technique (NFT), ebb and flow system (continuous flow culture), water culture, drip system, wick system, and aquaponic system. Similarly, aeroponics is also a soil-less method for plant production. Among all the currently available technologies that allow us to grow plants in a soil-less environment, aeroponics is the newest and most advanced. It is an advance version of hydroponics in which the plants are placed on a culture panel and the roots hang in air. Further, the nutrient solution is sprayed on the hanged roots with electronically controlled nutrient supply system.

ICAR-Central Potato Research Institute (CPRI), Shimla developed a state of the art aeroponic technology at Jalandhar with the use of IoT. It has huge potential to increase production in terms of number of mini-tubers per plant (40–70 mini-tubers per plant) from three to four times as compared to conventional method [2]. It facilitates congenial climatic conditions to produce targeted number of potato mini-tubers. The aeroponic unit consists of a grow box, solution tank, crop lifting system, and sensors and electronic controls. In this system, a series of nozzles are installed with the nutrient supply lines inside the grow box to supply the nutrients to the roots. A solution tank is provided with aeroponic unit to supply and store the nutrient solution for the plants. The aeroponic unit is provisioned with a PLC-based automatic nutrient supply system, which supplies prescribed amount of nutrients to the plant roots. An automatic EC and pH monitoring and dosing system are also integrated with the nutrient supply lines to maintain crop-specific pH and concentration of nutrient solution. Capacity of the solution tank depends upon the solution consumed by the plants per day.

Automatic controlled IoT-based greenhouse/polyhouse is also one of the advanced horticulture farming techniques. IoT played an important role in controlling the hardware unit remotely and automatically and also provided data to remote location. Some sensors (soil moisture, light, temperature, and humidity) and actuators are mounted inside the greenhouse and their data are sent by Wi-Fi module to the server. When the physical parameters reach their threshold value, micro-controller automatically actuates the motor, solenoid valve, exhaust fan and fogger to control the soil moisture or environment of greenhouse. Also, users access the data from Internet server and process them on different data analytical tools to estimate threshold value or predict the future values of physical parameters. Such type of system helps in controlling the biotic and abiotic stresses in greenhouse/polyhouse at an early stage.

Operational Issues and Challenges

During the past few years, various issues have emerged with the development of digital agriculture, which are related to proper maintenance, management, safety, and performance of smart devices implemented in agricultural practices [10].

Compatibility

Digital farming combines different technologies in a single platform; hence, it uses various protocols or standards for performing different tasks. Thus, computability issues arise when connecting different devices which worked on different firmware and operating systems. Sometimes compatibility problem arises when deploying software to hardware, like some controller boards are not capable for handling AI models.

Network Connectivity and Scalability

Connecting a large number of devices is one of the big challenges in digital world. The future of digital agriculture depends on the decentralization of networks. This issue has a greater impact on IoT devices, since IoT is completely based on internet connectivity. The proper availability of network is the basic requirement of IoT devices [6]. Network issues also occur when using Zigbees, LoRaWAN, Wi-Fi, and RFID technologies for communicating among different devices. Number of physical devices will be connected to the Internet in near future. So, the management of various devices over a variety of networks will be a threat. System scalability is another challenging and tedious work to achieve with collective consensus. This was more challenging when additional devices are added to the existing framework of software and hardware using different protocols.

Reliability

The system's reliability is the main goal of enhancing the success rate of the smart farming technologies. The system failure or virus threat is always a major challenge in hardware or software architecture of digital agriculture. Failure of network or smart devices can lead to loss of information, economic loss, and damage to crops. It is a very critical issue in autonomous robotic system because it is totally dependent on software and hardware system.

Data Confidentiality and Security

Possible cyber-attacks in digital services can lead to serious security issues like data integrity, data loss, data theft through smart applications, stealing of information in supply chain from stakeholders affecting agri-business, and unethical access of foreign sensors, drones, or robots in field. This data confidentiality and safety issues are arisen in the agriculture sector due to the internet connection of devices used in field. Therefore, there is a need to pay more attention to secure and complete data transmission in smart farming systems.

Intelligent Analysis and Actions

Analyzing field data and drawing good conclusions from the collected information is a major concern in AI-based robotic systems. The findings derived from data analysis tools help the autonomous system to take intelligent decisions. Hence, proper analysis of data is essential to get the right decision and accurate and precise action.

Lack of Skilled Workforce

For proper operation of smart devices requires skilled/experienced workers. Hence, training is required to end users for operating the software application or hardware systems with proper safety in field. A skilled workforce is not only required to handle software and hardware set up but also handling the issues related to agriculture practices like sowing or harvesting at right time, maintenance of electrical components or farm implements, etc.

Future Strategies for Smart Agricultural Mechanization

Present Indian agriculture is highly labor-intensive whereas smart agriculture is all about machines and technologies. The research institutes of the Indian Council of Agricultural Research (ICAR), SAUs, IITs, NITs, and other private organizations in India are involved in development of technologies based on precision agriculture, digital farming and AI through different projects such as the National Agricultural Innovation Project (NAIP), Consortia Research Platform on Farm Mechanization and Precision Farming (CRP on FMPF), AICRP on Farm Implements and Machinery, etc. These institutes are applying modern tools and techniques for application of sensors and robotics in planting, rice transplanting, spraying, weeding, drone-based spraying with the help of AI, etc. It is vital that these centers should focus on the stakeholders' interests to ensure that research concepts (farming methods and machinery) should not remain at the prototype stage [9]. The themes of PA, DF, and AI in agriculture can be applied across disciplines and may bring a paradigm shift in how we see farming today. The following strategies will not only enable farmers to do more with less but also help to improve quality and ensure faster go-to-market for crops.

1. Promotion of digital farming and precision agriculture technologies in addressing issues related to sustainable farming through research and development and financial assistance.
2. Lowering the cost of hardware/technology so that it is available and affordable for small and marginal farmers. The hardware/technology must be portable, plug-and-play type, and has better chance of success in India.

3. Establishment of an interactive digital platform to allow farmers full access to information and technology databases, expert systems, and DSS for web-based agro-advisory, skill development, machinery management, and financial assistance.
4. Need for increased application of smart agriculture and digital farming with involvement of private sector for farm mechanization.
5. The application of ground-based sensors and remote sensing data at high spatial and temporal scales can be integrated for forecasting and allocation of irrigation water.
6. Promotion of an app-based farmer-to-farmer aggregation platform, which bridges the demand and supply gap of machinery or equipment by connecting owners of tractors and farming equipment with those who require their services.
7. The drudgery prone and repetitive farm operations such as weeding, spraying, and harvesting of costly fruits and vegetables can be enabled with AI, leading to improved accuracy and productivity.

Conclusions

The smart digital technologies modernize the agriculture sector using IoT protocols, AI models, autonomous robotic system, UGV, and UAV. Smart agriculture utilizes the pipeline of ICT for sharing the information with experienced domain expertise to engender better solution of the problems related with farm practices. The smart farming techniques reduce environmental impact, hazards, and error; save time and cost; optimize inputs (water, fertilizer, and pesticides) usage, and increase crop yield and product quality. Digital agriculture is the hybridization of smart farming and precision agriculture. Digital agriculture collects, stores, processes the data, and develops actionable intelligent smart farming devices. Smart framing is applicable not only for large farms, but also for small and marginal lands. With the help of smart farming, farmers can plan and take decision through software-managed and sensor-enhanced system.

References

1. Alaba, F.A., Othman, M., Hashem, I.A.T., Alotaibi, F.: Internet of things security: a survey. J. Network Comput. Appl. **88**, 10–28 (2017)
2. Anonymous: Aeroponics for potato-seed production (2022). https://cpri.icar.gov.in//content/Index/?qlid=4071&Ls_is=4128&lngid=1
3. Ault, A.C., Krogmeier, J.V., Buckmaster, D.: Mobile, cloud-based farm management: a case study with trello on my farm. In: Kansas City, Missouri, July 21–July 24, 2013, p. 1. American Society of Agricultural and Biological Engineers (2013)
4. Bannerjee, G., Sarkar, U., Das, S., Ghosh, I.: Artificial intelligence in agriculture: a literature survey. Int. J. Sci. Res. Comput. Sci. Appl. Manage. Stud. **7**(3), 1–6 (2018)

5. Chandel, N.S., Chakraborty, S.K., Rajwade, Y.A., Dubey, K., Tiwari, M.K., Jat, D.: Identifying crop water stress using deep learning models. Neural Comput. Appl. **33**(10), 5353–5367 (2021)
6. Jindal, F., Jamar, R., Churi, P.: Future and challenges of internet of things. AIRCC's Int. J. Comput. Sci. Inf. Technol. **10**(2), 13–25 (2018)
7. Khanna, A., Kaur, S.: Evolution of Internet of Things (IoT) and its significant impact in the field of precision agriculture. Comput. Electron. Agric. **157**, 218–231 (2019)
8. Mehta, C.R., Chandel, N.S., Rajwade, Y.A.: Smart farm mechanization for sustainable Indian agriculture. Agric. Mechanization Asia Afr. Latin Am. **51**(4), 99–105 (2020)
9. Mehta, C.R., Tiwari, P.S., Chandel, N.S.: Smart farm mechanization—retrospective and prospective. Indian Farming **71**(11), 60–63 (2021)
10. Misra, N.N., Dixit, Y., Al-Mallahi, A., Bhullar, M.S., Upadhyay, R., Martynenko, A.: IoT, big data and artificial intelligence in agriculture and food industry. IEEE Internet Things J. **9**(9), 6305–6324 (2022)
11. Ozdogan, B., Gacar, A., Aktas, H.: Digital agriculture practices in the context of agriculture 4.0. J. Econ. Financ. Acc. **4**(2), 186–193 (2017)
12. Rahmadian, R., Widyartono, M.: Autonomous robotic in agriculture: a review. In: Third International Conference on Vocational Education and Electrical Engineering (ICVEE), pp. 1–6. IEEE (2020)
13. Tsouros, D.C., Bibi, S., Sarigiannidis, P.G.: A review on UAV-based applications for precision agriculture. Information **10**(11), 349 (2019)
14. Yaghoubi, S., Akbarzadeh, N.A., Bazargani, S.S., Bazargani, S.S., Bamizan, M., Asl, M.I.: Autonomous robots for agricultural tasks and farm assignment and future trends in agro robots. Int. J. Mech. Mechatron. Eng. **13**(3), 1–6 (2013)

Chapter 2
Applications of Geospatial and Big Data Technologies in Smart Farming

G. P. Obi Reddy, B. S. Dwivedi, and G. Ravindra Chary

Introduction

Globally, food production should be increased up to 70% by 2050 against the increasing population of 9.6 billion by 2050 [10]. On the other hand, the challenges from climate change in the form of increasing temperatures, weather variability, shifting agro-ecosystem boundaries, invasive pests and extreme weather events are affecting agriculture production [15]. In the context of pressure on land resources to enhance food production on one side and adverse impacts of climate change on the other, there is a need to adopt smart agriculture practices, where geospatial and big data technologies play an important role. Smart agriculture is a key tool to primarily handle and manage the threats, challenges and risks in the context of climate change, diseases and pest attacks and ensure sustainability. Smart farming practices include identification and localization of crops, insects and weeds; performance monitoring; machinery; variable rate of fertilizers, herbicides, insecticides and fungicides; planting monitoring and mapping [6]. Smart agriculture uses the knowledge of information science, environmental science, computer engineering, Geographic Information System (GIS), Global Positioning System (GPS), remote sensing technology and virtual satellite imaging for better integration with soil, climate, environment with agriculture [36]. In recent decades, due to the continuous development of Internet technology, various data explosions took place, among them, sensing technology and cloud computing technology play an important role in

G. P. Obi Reddy (✉) · B. S. Dwivedi
ICAR-National Bureau of Soil Survey and Land Use Planning, Amravati Road, Nagpur 440033, India
e-mail: GPO.Reddy@icar.gov.in

G. Ravindra Chary
ICAR-Central Research Institute for Dryland Agriculture, Santhosh Nagar, Hyderabad 500059, India

© Centre for Science and Technology of the Non-aligned and Other
Developing Countries 2023
K. Pakeerathan (ed.), *Smart Agriculture for Developing Nations*, Advanced Technologies and Societal Change, https://doi.org/10.1007/978-981-19-8738-0_2

smart farming. These two technologies revolutionized the amount of data generated, stored, analyzed and utilized on the basis of storage technology and cloud computing technology. The development of earth observation and sensing technology, especially satellite remote sensing, has made massive remotely sensed data available for research and various applications in smart farming [7, 20]. Spatiotemporal data derived from remote sensing are a valuable resource for smart farming to potentially generate voluminous big data and perform historical trends. Agricultural remote sensing is one of the backbone technologies for smart agriculture, which considers within-field variability for site-specific management instead of uniform management as in traditional agriculture.

Big data can be defined as the dataset that are too large, complex and exceeds traditional datasets for its acquisition, storage and management and analysis [47]. Big data is emerging as a potential technology for farm-level decisions to enhance farm productivity by detecting and overcoming cumbersome practices. Advantages of big data are high-speed transactions, provide access to large amounts of data and user can perform several operations simultaneously [34]. Smart agriculture uses the ICT, modern machinery equipment, Internet of Things (IoT), cloud computing, machine learning and big data analytics for crop's vegetative growth intercultural operation, climate-smart management, harvesting, post-harvest management and marketing management. In the IoT, wireless technologies play a central role in data gathering and data communication [1]. Wireless sensor networks (WSNs) and radio-frequency identification (RFID) are considered the two main building blocks of sensing and communication technologies for IoT [11, 22]. Wireless sensor networks' sensors, smart devices, RFID tags, tablets, palmtops, laptops, smart meters, smart phones, smart healthcare, social media, software applications and digital services generate the volume of data [22]. They continuously generate large amounts of structured, semi-structured and unstructured data, which are strongly increased in the field of smart agriculture. The combination of geospatial and big data applications in smart farming plays important role in various farm operations.

Geospatial Technologies and Their Components

Geospatial technologies comprise photogrammetry, satellite remote sensing, GIS and GPS, which have immense potential in the collection, analysis and interpretation of spatial data in the field of agriculture. These play a vital role in providing precise information on the nature, extent and spatial distribution of agricultural resources to assess their potential and limitations for planning, monitoring and management toward sustainable development [30, 32]. Remote sensing integrated with GIS provides an effective tool for the analysis of land resources at macro, meso and microlevels, which could potentially enhance the management of critical areas [35]. Several geospatial technologies are actively being used in the management of natural resources, including remote sensing from terrestrial, aerial and satellite platforms with various sensors, GPS and GIS [19]. The aerial and satellite remote sensing techniques by virtue of their speed and cost-effectiveness have an edge over conventional methods of the survey in mapping, monitoring and detecting the temporal changes in land degradation. High-resolution remotely sensed data provide an unparalleled view of the earth for studies that require synoptic or periodic observations such as inventory, surveying, mapping and monitoring in land resources, land use/land cover and environment. The integration of spatial data and their combined analysis could be performed through GIS and simple database query systems to complex analysis and decision support systems for effective land resource management. Agricultural remote sensing is a highly specialized field to generate images and spectral data in huge volume and extreme complexity to drive decisions for agricultural development. In the field of agriculture, conducted various remote sensing data-based studies in monitoring soil properties, crop stress and development of decision support systems in fertilization, irrigation and pest management to enhance crop production.

Remote Sensing

Information on the nature, extent and spatial distribution of various agricultural resources is a prerequisite for achieving sustainable agriculture. The advances in remote sensing technology provide data for detailed inventory and mapping of natural resources at different scales. A wide variety of satellite remote sensing data from Landsat OLI, IRS-P6, Cartosat-I, Cartosat-II, QuickBird, Sentinel-2 and Google are now available to the earth scientists for the generation of spatial databases on natural resources for various applications. Since remote sensing may not provide all the information needed for a full-fledged soil and land resource assessment, field survey and other ancillary information from various sources are needed to be integrated with remote sensing data. Various satellite sensors and their image characteristics are shown in Table 2.1.

Remote sensing technology allows us to observe the earth's features from space, and users can analyze the information collected from remote sensing of our interest. The advances in remote sensing technology offer data for detailed inventory, mapping and monitoring of agricultural resources at a large scale. A wide variety of satellite remote sensing data from MODIS, Landsat-ETM$^+$, IRS-IC, IRS-ID, IRS-P6, Sentinel, Cartosat-I Cartosat-II, QuickBird and Google are now available to the users for the generation of the spatial database on agricultural resources for various applications. These technologies are extensively used in land use/land cover changes, precision agriculture, yield estimation and other need-based agricultural applications. However, remotely sensed data coupled with field observations and contemporary technology provide more realistic datasets as compared to image interpretation alone. In this age of the twenty-first century with advancements in the computer field and in the field of geographic information systems, the resource management issues in smart agriculture are to be dealt with more efficiently, effectively and a convincing way by applying these geospatial technologies rather than the traditional methods

Table 2.1 Important multispectral and hyperspectral satellites, sensors and their image characteristics

	Satellite	Sensor	Year of launch	Spatial resolution (m)	Swath (kms)	Revisit (in days)
Multispectral	Terra	MODIS, ASTER, CERES, MOPITT, MISR	1999	250, 500 and 1000	2330	16
	Aqua	MODIS, AIRS, AMSR-E, AMER-U, CERES, HSB	2002	250, 500 and 1000	2330	16
	Landsat 8	OLI, TIRS	2013	15, 30 and 100	185	16
	Sentinel-2A	MSI	2015	10, 20 and 60	290	5
	Sentinel-2B	MSI	2017	10, 20 and 60	290	5
	Resourcesat-2A	AWiFS, LISS-3, LISS-4	2016	5.8, 23.5 and 56	370, 140 and 23.9	5 and 24
	Resourcesat-2	AWiFS, LISS-3, LISS-4	2011	5.8, 23.5 and 56	370, 140 and 23.9	5 and 24
	SPOT 6	NAOMI	2012	1.5	10	1
	SPOT 7	NAOMI	2014	1.5	10	1
	CartoSat-2B	PAN	2012	0.8	9.6	4
	GeoEye-1	GIS-MS, GIS-PAN	2008	0.41 and 1.65	15.2	
	WorldView-3	CAVIS, MSS, PAN, SWIR	2014	0.31, 1.24, 3.7 and 30	13.1 and 13.2	1
	CartoSat-3	MX, PAN	2019	0.28	16	4
Hyperspectral	GEO-KOMPSAT-2B	GEMS, GOCI-II	2020	250 and 7000	2500	
	HysIS	HySI	2018	30	30	
	PRISMA	PRISMA	2019	5 and 30	30	29
	EnMAP	HSI	2020	30	30	27
	HISUI	HISUI	2019	20	20	

Source Reddy and Kumar [29]

of management. In land resources' applications, there are a wide range of satellite platforms providing imagery at multiple resolutions with frequency at the global to local scale. Microwave remote sensing is highly useful, as it provides observation of the earth's surface, regardless of day/night and atmospheric conditions. Space-borne microwave remote sensors provide perspectives of the earth's surface and atmosphere, which are of unique value in studies of geomorphology, topography and vegetation classification. Hyperspectral remote sensing imagers acquire many, very narrow, contiguous spectral bands throughout the visible, near-infrared, mid-infrared and thermal infrared portions of the electromagnetic spectrum enabling the construction of an almost continuous reflectance spectrum for every pixel in the scene.

Geographic Information System (GIS)

GIS is defined as a powerful set of tools for collecting, storing, retrieving at will, displacing and transforming spatial data [5, 33]. GIS is a computer system capable of capturing, storing, analyzing and displaying geographically referenced information, i.e., data identified according to location. A true GIS is designed to accept, organize, statistically analyze and display diverse types of spatial information that is geographically referenced to a common coordinate system of a particular projection and scale. GIS is a potential tool for handling voluminous remotely sensed data and can support spatial statistical analysis, presentation of spatial data in the form of a map, as well as storage, management, modeling of input data and presentation of model results [31]. Data derived from the latest remote sensing and information technology techniques could be effectively integrated in GIS to perform the analysis. Applications of GIS range from simple database query systems to complex analysis and decision support systems. GIS techniques are playing an increasing role in facilitating the integration of multilayer spatial information with statistical attribute data in the sustainable management of agricultural resources.

Global Positioning System (GPS)

GPS has revolutionized research in the areas of surveying, engineering, monitoring positions and navigation [28, 45]. GPS technology enables real-time data collection and accurate position information, which in turn leads to efficient analysis and manipulation of large amounts of geospatial data in smart farming. For mapping, where surveying accuracy better than a meter is required, specialized DGPS techniques have been developed. These techniques are being employed increasingly for detailed topographic, soil, geological, engineering and environmental-related surveys at different scales. GPS-based surveying systems allow surveys to increase accuracy in mapping over conventional surveying techniques. GPS plays a significant role in field data collection, surveying and monitoring the crops on a periodical basis. GPS provides current position, anywhere and at any time on the globe, with a reasonable degree of accuracy. In the field of smart agriculture, GPS technology can help to map the relief, farm boundaries and assets more precisely, which serves as a "baseline" for further farm decision activities. Tractor-mounted GPS with variable-rate applicators' farmers can identify locations that are nutrient deficient and apply the appropriate amounts of fertilizer in specific locations to enhance the fertilizer efficiency and improve the crop production in smart agriculture.

Big Data and Its Components

Big data is described as a collection of data that is extremely large and cannot be collected, processed, stored and calculated within the time required by traditional data processing methods or tools. Big data represents the information assets characterized by high volume, velocity, variety, veracity, variability and value to require specific technology and analytical methods for its transformation into value. Big data is being used to provide predictive insights in smart farming operations, drive real-time operational decisions and redesign business processes for game-changing farm operations. Big data consists of large datasets (*volume*), which have different types of datasets. *Velocity* consists of the speed by which information moves through the system, where data flow into the system from multiple sources. A *variety* of data can be found in big data, which is often unique because of a wide range of both the sources being processed and their relative quality. *Veracity* helps to evaluate the quality of data and processing quality of result analysis. *Variability* includes variation in the data that leads to wide variation in quality, where additional resources can be used to identify, process or filter low-quality data to make it more useful. To measure the *value* of data, the big data ecosystem deals with boundless processes and approaches both in structured or unstructured data. Agricultural big data is mainly used in agricultural condition monitoring, agricultural product monitoring and early warning, precision agricultural decision-making and the construction of a rural comprehensive information service system [14, 40].

Big data is considered a potential technology for the assessment of farm-level decisions, policy decisions and market-distorting actions for increasing the productivity in agriculture properly. Big data technology in agriculture can collect and analyze lots of data, which are usually generated from various sectors and stages in agriculture. Big data-driven agriculture provides an opportunity to transform traditional decision-making to database decision-making. Cloud computing platforms are efficient ways of storing, accessing and analyzing datasets on very powerful servers, which virtualizes supercomputers for the user. These systems provide infrastructure, platform, storage services and software packages in a variety of ways for the customers [7, 21]. Google earth engine (GEE) is a big data cloud computing platform, and it facilitates scientific discovery processes with free access to numerous remotely sensed datasets to perform various geo-big data analytics in the field of agricultural applications [12,

Fig. 2.1 Potential
applications of GEE cloud
computing platform in smart
farming

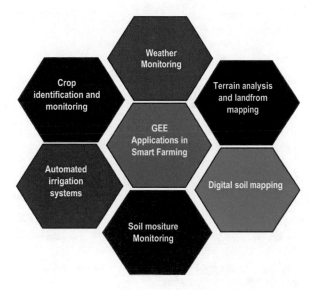

39]. GEE provides various functions to perform spectral and spatial operations on either a single image or a time-series images. Different operations within the GEE platform range from simple mathematical operations to advanced image processing and machine learning algorithms. The important GEE applications in smart agriculture are real-time weather monitoring, terrain analysis and landform mapping, digital soil mapping, soil moisture monitoring, automated irrigation systems, crop identification and monitoring, yield monitoring and mapping, pest monitoring and surveillance and variable-rate technology (VRT) which are briefly summarized (Fig. 2.1).

Applications of Geospatial and Big Data Technologies in Smart Farming

In smart farming, geospatial technologies like high-resolution remote sensing, GIS, GPS and field sensors play an important role in the reorganization of spatial variations at individual farm level and to address those using appropriate strategies.

Weather Monitoring

One of the biggest challenges for smart farming is climate change and its impact on crop productivity and farm operations. In smart farming, farmers essentially need accurate information on various weather parameters like rainfall, temperature, humidity and weather forecasts. A wide range of ground, aerial and space-based earth

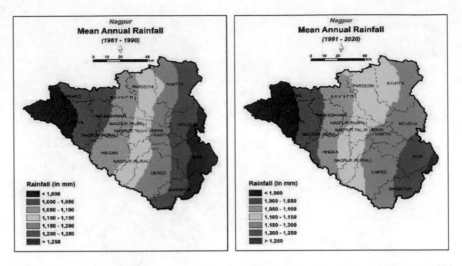

Fig. 2.2 Mean annual rainfall pattern of Nagpur district, Maharashtra, India, during 1961–1990 and 1991–2020

observation platforms are available to collectively enable real-time weather observation and dynamic monitoring of accurate, comprehensive, continuous and diverse information on weather parameters. Automatic weather station (AWS) provides information on various meteorological parameters, which include rainfall, temperature, humidity and soil moisture for better-informed decisions in smart farming. By collecting the data on time-series various weather parameters, one can analyze the climatic trends by using data analytics tools in GIS to assess the climatic conditions for smart farming. As an example, the mean annual trends' rainfall of Nagpur district of Maharashtra, India, was analyzed for the period from 1960 to 1990 and 1991 to 2020. The analysis shows that there is a slight change in the mean annual rainfall pattern in the district over a period of 60 years. The mean annual rainfall in the eastern regions of the district shows a decreasing trend, and it has an impact on agriculture and crop production (Fig. 2.2).

Terrain Analysis and Landforms Mapping

Integration of satellite-based remote sensing with photogrammetry, GIS and GPS has enhanced its capabilities in the area of agricultural resource management and solving environmental problems very rapidly and efficiently in comparison to others [18]. Remote sensing integrated with GIS provides an effective tool for the analysis of land resources at macro, meso and microlevels, which could potentially enhance the management of critical areas [35]. High-resolution remotely sensed data provide an unparalleled view of the earth for studies that require synoptic or periodic observations such as inventory, surveying, mapping and monitoring of land resources,

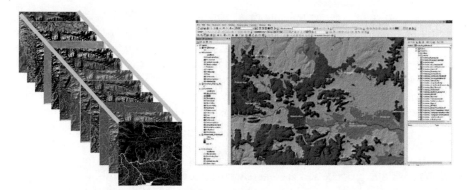

Fig. 2.3 Delineation of landforms through digital terrain modeling in basaltic terrain of Central India

land use/land cover and environment. Traditional mapping approaches are qualitative interpretation with inherent limitations associated with fieldwork and human interventions and domain knowledge. Nowadays, with access to fast computers and digital sources such as digital elevation models (DEMs) acquired by remote sensing, landforms can be analyzed and mapped digitally. In digital terrain parameterization, various attributes such as elevation, slope, aspect, plan and profile curvature and flow accumulation were analyzed and mapped [24] to obtain distinct landforms units. Relatively recent advances in geospatial technologies and developments in numerical modeling of surface processes have revolutionized the field of geomorphology. The digital terrain modeling approaches help to quantify landscape morphology [13, 26], assess surface biophysical conditions [38], landform process [2] and improve the landform mapping accuracy. The automatically extracted landforms from digital terrain modeling of SRTM (30 m) data for basaltic terrain of Central India are shown in Fig. 2.3.

Digital Soil Mapping

Intimate knowledge of the kind of soils and their spatial distribution is a prerequisite in developing site-specific land-use plans in smart farming, irrigation scheduling, drainage management, etc. Space technology plays a significant role in digital soil mapping efforts. Digital soil mapping involves the quantitative prediction of soil properties by using some observed soil data and some soil-forming factors [8]. Advances in digital soil mapping have created a tremendous potential and transformed the procedure and the way the soil maps are produced [23]. However, in general, soils appear to be more continuous variables than discrete objects [27]. Therefore, the conceptualization of various terrain variables as discrete objects involves uncertainties. Such variables are best predictors through the fuzzy logic approach and minimize the uncertainties. The application of fuzzy logic models in digital soil mapping requires the establishment of knowledge bases to define the fuzzy membership criteria. For digital soil mapping, the fuzzy logic approach uses the principle that a spatial unit, e.g., a pixel, can contain soil, which cannot be exclusively classified into one soil class. Membership values in fuzzy mapping can be determined using deterministic or empirical approaches. Digital soil mapping begins with the development of a numerical or statistical model of the relationship between environmental variables and soil properties, which is then applied to a geographic database to create a predictive map. Digital soil mapping helps to exploit the relationship between environmental variables and soil properties in order to more efficiently collect soil data, produce and present data that better represent soil-landscape continuity and explicitly incorporate expert knowledge in model design. Soil parameters' map developed through digital soil mapping with raster grid allows better characterization of soil-landscape variability, which in turn significantly helps to formulate land-use plans in smart farming based on the spatial variability of soil characteristics.

Soil Moisture Monitoring

In smart farming, soil moisture is a fundamental factor that is used to control the water management system for sustained crop yield. Agricultural UAV or drone platforms become one of the most promising technologies utilized in soil moisture monitoring. The agricultural UAV with a combination of IoT sensors can detect moisture conditions of the soil and helps in the identification of zones that suffer from water scarcity and dryness of soil profile [37]. Such information helps the farmers to take appropriate interventions such as irrigation scheduling and amount of water to be released. High-resolution satellite imageries can be effectively used for monitoring soil moisture and assessment of surface moisture conditions in real time across the entire surface, rather than for a single point like the soil moisture, and weather sensors provide soil moisture. It is important to build time-series soil moisture data for individual fields to interpret and understand the spatial patterns as soil type and texture affect the soil physics of moisture and storage. Real-time monitoring of soil moisture helps to generate timely soil moisture information and inform crop management decisions, such as fertilizer inputs, yield mapping and pest risk assessment in smart farming.

Automated Irrigation Systems

An automated irrigation system does the work without the involvement of many manual operations by using computers and sensors. A rule-based expert system helps to save the water a lot, where field sensors collect and analyze data by an automated program, and accordingly, required water is supplied to the required part of the farm. An IoT-enabled, remote-controlled cross-platform system equipped with big data analytics and intelligent irrigation scheduling can be accessed through web and mobile applications, which are hosted on the enterprise cloud. These systems control the operation and help in crop field assessment for irrigation and applying fertilizers and pesticides. An automated irrigation system using WSN and General packet radio service (GPRS) module could be used to optimize the use of water for crops [16]. This system works based on the distributed wireless sensor network with soil moisture and temperature sensor in WSN. Gateway systems could be used to transfer data from the sensor unit to the base station, send commands to the actuator for irrigation control and manage data of the sensor unit. The traditional techniques like drip irrigation and sprinkler irrigation adopted in smart farming can be combined with IoT-based systems to improve water-use efficiency. IoT helps to access information and make major decision-making processes by getting different values from sensors like soil moisture, water level sensors and water quality.

Crop Identification and Monitoring

Geospatial and big data technologies enable us to identify and monitor crop conditions on a periodical basis. High-resolution satellites and drones are capable of collecting spatial data on various parameters of crops that help in analyzing and monitoring the current crop health. Drones equipped with various types of high-frequency sensors are capable of capturing and transmitting the data on the go. Drones can identify crop growth by comparing time-series images taken by the satellite and can assist farmers to reduce excessive use of water and reduce the chemical load on the environment by spraying required pesticides on the plant require. Object-based image analysis techniques [3, 44] provide an innovative approach to perform image segmentation with similar spectral signatures into objects to classify. IoT-based crop monitoring provides a data-driven approach with the use of sensor devices, gateway connectivity and a user-friendly dashboard to identify and monitor the crops in real time.

Yield Monitoring and Mapping

In fully mechanized farming systems, grain yield monitors are continuously monitored and measured in the clean-grain elevator of a combine. When systems are coupled with a GPS receiver, yield maps can be prepared with the help of yield monitors, which significantly helps in making sound management strategies. In yield mapping, soil, landscape and other environmental factors should also be weighed. The yield information provides insights into determining the effects of managed inputs. Various metrological factors like precipitation, temperature, humidity and solar light were used for crop yield maximization and to improve the quality of crops by considering real-time data and ML algorithms [25]. GPS sensors can be mounted on harvesting equipment to fetch the spatial coordinates and generate a crop yield map.

Pest Monitoring and Surveillance

Recent advances in remote sensing technology and geospatial image processing using drones have enabled the rapid monitoring, detecting and surveillance of insect pests and diseases in smart farming [42]. For detection of the occurrence and monitoring of pests and diseases on farms, remote-sensing technologies like satellites and drones can be employed to find insect pests and inform farmers of the state of affairs promptly [46]. In smart farming, an automatic insect identification system with various sensors can be built for insect and pest detection on crops through image analysis. Low-altitude remote-sensing technology like drones has the characteristics of high flexibility and image definition which can be deployed in pest and disease monitoring for crops. The deep learning and image process models enable us to identify any crop diseases or pest infestation within the crops [43]. The application of deep learning algorithms is an innovative method for image processing and object detection in the classification of various crop diseases [17]. Smartphone-based AI-powered applications could alert farmers and expedite disease diagnosis, potentially preventing or limiting pest and disease outbreaks. The IoT technology used to monitor pests and diseases has not yet fully developed as a unified standard in terms of services and equipment development.

Variable-Rate Technology (VRT)

Variable-rate technology (VRT) is an integral part of smart farming that allows data to be used effectively both in time and space. It combines the necessary farm equipment, control mechanisms and software tools to apply required quantities of inputs at specific times or locations to optimize fertilizer-use efficiency. Farm inputs like fertilizers, agrochemicals and irrigation water can be applied and monitored at different rates across a field, without manually changing rate settings on equipment. By utilizing geospatial technologies like GIS, GPS and field variability maps, VRT helps to increase the input efficiency, minimize overapplication of inputs and reduce the risk of pesticide and fertilizer runoff or leaching into water sources. Tractor-mounted sensor-based VRT utilizes sensors effectively to assess crop or field variability to provide real-time variable-rate application (VRA) of inputs.

Geospatial and Big Data Technologies in Smart Farming—Challenges

In the field of geospatial and big data technologies, the acquisition and creation of valuable data need strong hardware and software infrastructure for seamless data collection, quality, storage and integration. It would need high-performance computing for subsequent data processing, analysis and effective management of big data. Availability of reliable, timely and quality of spatial and non-spatial data is often a challenge in smart farming applications. Interoperability becomes the most important point of concern in a seamless flow of information [9]. This involves a number of issues like technical, semantic, syntactic and organization interoperability to be addressed [4, 41]. Designing high-performance systems is essential for easy access to distributed data by different users. Big spatial data computing requires modern computing and analytical methods to analyze the unevenly distributed data originating in real time from different locations. The innovative open-source cloud platforms like GEE can be used to access services to analyze voluminous data. In data-driven smart farming, data privacy issues and rights to data usage are quite obvious. The lack of data governance and suitable policies in place hinders data privacy and security. Hence, privileged access to big data and building trust with farmers should be a starting point in developing robust applications in smart farming. Since big data is from various sources in the agriculture supply chain, multistakeholder collaboration is needed for improved decisions and quick access to the right data to evaluate key performance indicators in building successful applications.

Geospatial and Big Data Technologies in Smart Farming—Opportunities

Some of the emerging opportunities of geospatial and big data applications in smart farming are to design and develop GPS-enabled tractors. These systems help to find out soil quality and quantity of crops from different areas of the field and determine how much fertilizer is to be applied in each part of the field. Such systems help the farmers to accurately navigate to specific locations within the field to collect soil samples or monitor crop conditions. As per requirement, different sensors can be established in the field to acquire the required data in real time. An environmental monitoring system can be implemented for real-time measuring temperature, humidity, illumination and soil moisture by using an array of sensors. Smart greenhouse helps to monitor temperature, humidity and soil moisture to change the climate conditions according to the requirement of plants for effective growth. These smart greenhouses minimize the human efforts in handling the operations. There is an ever-increasing awareness of the necessity to develop and apply robotic systems in agriculture, forestry, greenhouses, horticulture, etc. Smart robotic vehicles are capable of working round the year in all weather conditions and have the intelligence embedded within them to behave sensibly in a semi-natural environment, unattended while carrying out many useful tasks in smart farming. In smart farming, farmers take final decisions using various strategies and different services. In this process, all the decisions are based on the previous result and different algorithms come into the action to analyze the previous results and enable the user to take the final decision in smart farming operations. In the supply chain, perishable and sensitive materials such as seeds, plants and food products prevent spoilage which is a matter of concern, where big data has proven useful at various stages. At the production stage, automated systems handle data to show performance and reveal issues in critical equipment. Big data helps farmers and suppliers optimize fleet management to increase delivery reliability. Big data tracking solutions, smart meters and GPS-oriented analytics improve routing, cutting transportation costs and offering advanced mapping of the locations. The other common challenges are the big data life cycle in different applications, such as the appropriate data identification, data deployment, data representation, data fusion, as well as data visualization and interpretation.

A Way Forward

The convergence of multiple technologies such as geospatial, big data, AI, IoT and cloud computing has a significant impact on various operations and applications in smart farming. In smart farming, there are several processes, which demand higher energy input, and markets require products of high quality. Such a process can be classified according to the requirement, technology and applications. AI can augment automation and cost reduction and enhance revenue. ML algorithms can overcome human limitations in terms of speed, accuracy, reliability, consistency and transparency in performing the assigned tasks. Robotics associated with AI need intelligence to handle tasks like object manipulation and navigation in conjunction with subproblems of localization, motion planning and mapping. IoT is used mainly for the automation of equipment and application development using data analytics. In IoT-based smart farming, monitoring crop fields with the help of sensors and automating the irrigation systems are possible, where farmers can monitor and control the farm conditions from a remote. IoT enables farmers to take a data-driven approach to collect vast amounts of information from instrumented sensors about the status of their farms to improve farm yield and mitigate risks from weeds, pests and diseases. The advanced sensing technologies that can be deployed for smart farming include proximal, airborne and satellite-based sensors. The use of mobile and smart phone technologies in smart farming could be effectively used in harnessing the full potential in the communication space. Several smart phone applications are available for agriculture, horticulture and farm machinery to make farm operations smarter and easier.

Conclusions

The integration of remote sensing data with GIS, machine learning and deep learning offers great potential for better extraction of geographical information from remote sensing data and images. With the increased volume and complexity of remote sensing data acquired from multiple sensors using multispectral and hyperspectral devices with multiangle views with time, new development is needed for visualization tools with spatial, spectral and temporal analysis. Temporal satellite data have become valuable tools in studying the spatial extent of degraded lands and for mapping and monitoring the changes that have taken place over a period of time due to reclamation/conservation measures. Smart farming powered by geospatial and trending technologies like AI with big data analytics, IoT, AI, ML and DL immensely helps in making farming smarter with good profit margins. In data-driven smart farming, big data access, the availability of quality data, big spatial data computation, spatial data integration, data privacy and rights to use data are some of the important challenges. On the other hand, these technologies offer immense opportunities in smart farming. Farmers can access the facilities of the greenhouse through the dashboard, and operations can be managed by using voice commands. In a smart farming supply chain, the use of big data is found to be immensely useful at all stages. The geospatial technologies in conjunction with data science techniques have immense potential in spatiotemporal analysis of land resources and land transformation processes, which in turn helps to plan and manage them for optimum utilization of land resources to achieve sustained production levels and food security.

References

1. Abdmeziem, M.R., Tandjaoui, D., Romdhani, I.: Architecting the internet of things: state of the art. Robots Sens. Clouds 55–75 (2016)
2. Allen, T.R., Walsh, S.J.: Characterizing multitemporal alpine snowmelt patterns for ecological inferences. Photogramm Eng. Remote Sens. **59**(10), 1521–1529 (1993)
3. Blaschke, T.: Object-based image analysis for remote sensing. ISPRS J. Photogramm Remote Sens. **65**, 2–16 (2010)
4. Brewster, C., Roussaki, I., Kalatzis, N., Doolin, K., Ellis, K.: IoT in agriculture: designing a Europe-wide large-scale pilot. IEEE Commun. Mag. **55**(9), 26–33 (2017)
5. Burrough, P.A., McDonnell, R.A.: Principles of Geographical Information Systems. Oxford University Press, Oxford (1998)
6. Buyya, R., Dastjerdi, A.V.: Internet of Things: Principles and Paradigms. Elsevier, New York (2016)
7. Chi, M., Plaza, A., Benediktsson, J.A., Sun, Z., Shen, J., Zhu, Y.: Big data for remote sensing: challenges and opportunities. Proc. IEEE **104**, 2207–2219 (2016)
8. Dobos, E., Carré, F., Hengl, T., Reuter, H.I., Tóth, G.: Digital soil mapping as a support to production of functional maps. Office for Official Publications of the European Communities, Luxemborg. EUR, 22123, 68 (2006)
9. Fakhruddin, H.: Precision agriculture: top 15 challenges and issues (2020). https://plagiaris mdetector.net/teks.co.in/site/blog/precision-agriculture-top-5challenges-and-issues
10. FAO: E-agriculture in Action. Italy, Rome (2017)

11. Fortino, G., Savaglio, C., Spezzano, G., Zhou, M.: Internet of things as system of systems: a review of methodologies, frameworks, platforms, and tools. IEEE Trans. Syst. Man Cybern.: Syst. (2020)
12. Gorelick, N., Hancher, M., Dixon, M., Ilyushchenko, S., Thau, D., Moore, R.: Google earth engine: planetary-scale geospatial analysis for everyone. Remote Sens. Environ. **202**, 18–27 (2017)
13. Hengl, T., Reuter, H.I. (eds.): Geomorphometry: Concepts, Software, and Applications. Developments in Soil Science. Elsevier, Amsterdam (2009)
14. Hou, L., Wang, X.D., Gao, Q., et al.: Construction of agricultural big data mining system based on Hadoop. J. Libr. Inf. Sci. Agric. **30**(7), 19–21 (2018)
15. IPCC: Climate Change and Land: an IPCC Special Report on Climate Change, Desertification, Land Degradation, Sustainable Land Management, Food Security, and Greenhouse Gas Fluxes in Terrestrial Ecosystems, p 864 (2019)
16. Jaguey, J.G., Villa-Medina, J.F., Lopez-Guzman, A., Porta-Gandara, M.A.: Smartphone irrigation sensor. IEEE Sens. J. **15**, 5122–5127 (2015)
17. Kamilaris, A., Prenafeta-Boldú, F.X.: Deep learning in agriculture: a survey. Comput. Electron. Agric. **147**, 70–90 (2018)
18. Kanniah, K.D., Hashim, M.: A systematic approach in remote sensing education and training in Malaysia (with Special reference to Universiti Teknology Malaysia). Int. Arch. Photogramm. Remote Sens. **33**(B6), 153–163 (2000)
19. Kingsford, R.T.: Managing the water of the Border Rivers in Australia: irrigation, government and the wetland environment. Wetland. Ecol. Manag. **7**(1), 25–35 (1999)
20. Liu, P.: A survey of remote-sensing big data. Front. Environ. Sci. **3**, 1–6 (2015)
21. Ma, Y., et al.: Remote sensing big data computing: challenges and opportunities. Future Gener. Comput. Syst. **51**, 47–60 (2015)
22. Marjani, M., Nasaruddin, F., Gani, A., Karim, A., Hashem, I.A.T., Siddiqa, A., Yaqoob, I.: Big IOT data analytics: architecture, opportunities, and open research challenges. IEEE Access **5**, 5247–5261 (2017)
23. McKenzie, N.J., Jacquier, D., Ashton, L.J., Cresswell, H.P.: Estimating soil properties using the Atlas of Australian Soils. Technical Report 11/00, CSIRO Land and Water, Canberra (2000)
24. Moore, I.D., Lewis, A., Gallant, J.C.: Terrain attributes: estimation methods and scale effects. In: Jakeman, A.J., Beck, M.B., McAleer, M.J. (eds.) Modeling Change in Environmental Systems, pp. 189–214. Wiley, New York (1993)
25. Mulge, M., Sharnappa, M., Sultanpure, A., Sajjan, D., Kamani, M.: An invitation to subscribe. Int. J. Analy. Experiml. Modal. Analy. **10**(1), 1112–1117 (2020)
26. Pike, R.J.: Geomorphometry: diversity in quantitative surface analysis. Prog. Phy. Geogr. **24**, 1–20 (2000)
27. Qi, F., Zhu, A.-X., Harrower, M., Burt, J.E.: Fuzzy soil mapping based on prototype category theory. Geoderma **136**, 774–787 (2006)
28. Reddy, G.P.O.: Global positioning system: principles and applications. In: Reddy, G.P.O., Singh, S.K. (eds.) Geospatial Technologies in Land Resources Mapping, Monitoring and Management. Geotechnologies and the Environment, vol. 21, pp. 63–74. Springer, Cham (2018c)
29. Reddy, G.P.O., Kumar, K.C.A.: Machine learning algorithms for optical remote sensing data classification and analysis. In: Reddy, G.P.O., et al. (eds.) Data Science in Agriculture and Natural Resource Management, vol. 96, pp. 195–220. Springer (2022)
30. Reddy, G.P.O., Patil, N.G., Chaturvedi, A.: Sustainable Management of Land Resources—an Indian Perspective, pp. 796. Apple Academic Press Inc., Canada (2017)
31. Reddy, G.P.O.: Spatial data management, analysis, and modeling in GIS: principles and applications. In: Reddy, G.P.O., Singh, S.K. (eds.) Geospatial Technologies in Land Resources Mapping, Monitoring and Management. Geotechnologies and the Environment, vol. 21, pp. 127–142. Springer, Cham (2018b)
32. Reddy, G.P.O., Singh, S.K.: Geospatial Technologies in Land Resources Mapping, Monitoring, and Management, Geotechnologies and the Environment, vol. 21, pp. 638. Springer (2018)

33. Reddy, G.P.O.: Geographic information system: principles and applications. In: Reddy, G.P.O., Singh, S.K. (eds.) Geospatial Technologies in Land Resources Mapping, Monitoring and Management. Geotechnologies and the Environment, vol. 21, pp. 45–62. Springer, Cham (2018a)
34. Reddy, G.P.O., Dwivedi, B.S., Chary, G.R.: Big data in smart farming: challenges and opportunities. Indian Farming 71(11), 75–78 (2021)
35. Reddy, G.P.O., Maji, A.K., Nagaraju, M.S.S., Thayalan, S., Ramamurthy, V.: Ecological evaluation of land resources and land-use systems for sustainable development at watershed level in different agro-ecological zones of Vidarbha region. In: Maharashtra using Remote sensing and GIS Techniques, Project Report, NBSS & LUP, Nagpur, 270p (2008)
36. Schuster, J.: Big data ethics and the digital age of agriculture. Resour. Eng. Technol. Sustain. World 24(1), 20–21 (2017)
37. Slalmi, A., Chaibi, H., Saadane, R., Chehri, A., Jeon, G., Aroussi, H.K.: Energy-efficient and self-organizing internet of things networks for soil monitoring in smart farming. Comput. Elect. Eng. 92, e107142 (2021)
38. Smith, M., Pain, C.: Applications of remote sensing in geomorphology. Prog. Phy. Geogr. 33, 568–582 (2009)
39. Tamiminia, H., Salehi, B., Mahdianpari, M., Quackenbush, L., Adeli, S., Brisco, B.: Google earth engine for geo-big data applications: a meta-analysis and systematic review. ISPRS J. Photogramm Remote Sens. 164, 152–170 (2020)
40. Tao, Z.L., Guan, X.F., Chen, Y.W.: Construction of information sharing platform based on agricultural big data. Ind. Technol. Forum 17(11), 56–57 (2018)
41. Tayur, V.M., Suchithra, R.: Review of interoperability approaches in application layer of internet of things. In: International Conference on Innovative Mechanisms for Industry Applications (ICIMIA), pp. 322–326. IEEE (2017)
42. Vanegas, F., Bratanov, D., Powell, K., Weiss, J., Gonzalez, F.: A novel methodology for improving plant pest surveillance in vineyards and crops using UAV-based hyperspectral and spatial data. Sensors 18, 260 (2018)
43. Vijayakanthan, G., Kokul, T., Pakeerathai, S., Pinidiyaarachchi, U.A.J.: Classification of vegetable plant pests using deep transfer learning. In: 10th International Conference on Information and Automation for Sustainability (ICIAfS), pp. 167–172 (2021). https://doi.org/10.1109/ICIAfS52090.2021.9606176
44. Walter, V.: Object-based classification of remote sensing data for change detection. J. Photogramm Remote Sens. 58, 225–238 (2004)
45. Xu, S., Zhang, H., Yang, Z.: GPS Measuring Principle and Application, 3rd edn., pp. 1–10. Wuhan University of Technology Press, Wuhan (2008)
46. Zheng, Q., Huang, W., Cui, X., Shi, Y., Liu, L.: New spectral index for detecting wheat yellow rust using sentinel-2 multispectral imagery. Sensors 18, 868 (2018)
47. Zhou, X.C., Chen, Y.M., Zhu, X.H.: A kind of agricultural internet of things big data platform architecture. Anhui Agric. Sci. 47(2), 241–245 (2019)

Chapter 3
Applications of Drones in Smart Agriculture

Satya Prakash Kumar, A. Subeesh, Bikram Jyoti, and C. R. Mehta

Introduction

The growing population and increasing food demand have put immense pressure on the agriculture sector to increase productivity. Issues induced by climate change and other factors have also introduced further hurdles; thus, the adoption of innovative ways is the only option that remains. It has been proven that coupling digital technologies with existing farming practices is a practical approach to combat and fight against climate change, labor shortage, water scarcity, etc. Technology-driven precision farming practices that help farmers to take better informed decisions have to be given more emphasis at this point. Precision agriculture has been introduced to achieve more productivity, profitability, and environmental safety by controlling the input level in a site-specific manner [43]. Earlier, in the 2000s, satellite imagery was the main source of information, providing insights into the fields. The main disadvantage of satellite imagery in agriculture was the poor resolution, low frequency in time, and considerable period between observation and usage [31]. Aerial imagery from balloons and aircraft was further options employed for data collection. However, the high cost involved in the data collection was a major drawback. Drone-based imagery in agriculture has been a major breakthrough as it gives large coverage area, immediate and mosaic images within a short span of time that can be used for high-quality decision-making.

A drone is an unmanned aircraft that can be controlled remotely by an operator and carries multiple sensors, cameras, etc., through which aerial images are captured. Drones have been extensively used in commercial industries for the past few decades. But it started proliferating in the last few years into more practical applications.

S. P. Kumar (✉) · A. Subeesh · B. Jyoti · C. R. Mehta
ICAR-Central Institute of Agricultural Engineering, Bhopal, India
e-mail: satyaprakashkumar27@gmail.com

© Centre for Science and Technology of the Non-aligned and Other
Developing Countries 2023
K. Pakeerathan (ed.), *Smart Agriculture for Developing Nations*, Advanced Technologies
and Societal Change, https://doi.org/10.1007/978-981-19-8738-0_3

Owing to the advancement of information and communication technology, artificial intelligence, and Internet of Things, today drones are widely employed across different industries, including agriculture, enabling quick, high-quality, and cost-effective operations. Drones in agriculture have opened up new horizons to increase the agriculture outputs with enhanced farming applications and real-time access to high-quality information. Crop monitoring has become a simple task with the emergence of drone-based data collection, replacing the traditional labor-intensive and time-consuming data collection methods. While drones have just been into the mainstream agriculture domain, they are playing a crucial role in precision farming, helping farmers in achieving sustainable farming practices along with increasing profitability. The most common application in precision agriculture is to assess the health of the vegetation using remote sensing and image analytics. Besides the crop monitoring, another potential application is spraying on crops. This process has been initiated in Japan by installing small tanks in unscrewed helicopters. Nowadays, UAVs can carry huge tanks with a capacity of more than 10 L and make it possible to cover a larger area in very short time [35]. Integrating the drone with advanced techniques like artificial intelligence and deep learning improves the quality of solution being developed. When the real-time deep learning-based image analysis is coupled with UAV imagery, intelligent target spraying can be achieved. The use of drones in agriculture domain is steadily growing as it assists farmers in streamlining operations, making use of advanced data analytics and gaining hidden insights about their farms.

Definition and Types

Essentially, a drone is defined as a remotely piloted aircraft controlled by a human operator via a radio link or with various levels of autonomy achieved by using autopilot technology [46]. Drones are more commonly known as unmanned aerial vehicles (UAVs) or unmanned aircraft systems (UASs) or remotely piloted aircraft (RPA).

In general, drones are categorized into three categories based on their aerodynamic characteristics, namely fixed-wing, rotary-wing, and hybrid. Each type of drone has its pros and cons, especially concerning takeoff and landing requirements, flight time, payload capacity, coverage area, etc. Various types of drone models have been introduced in the past few years. Rotary wings, especially multirotors, are the most common types of drone used by professionals and hobbyists alike for aerial photography and surveillance. Based on the number of rotors, a rotary-wing drone can be classified as tri-copters (three rotors), quad-copters (four rotors), hexa-copters (six rotors), and octa-copters (eight rotors) respectively [18]. In India, drones are commonly categorized in five types based on their weight, i.e., nano (< 0.25 kg), micro (0.25–2.0 kg), small (2.0–25 kg), medium (25–150 kg), and large (> 150 kg) [4].

Drone Hardware

The main hardware components of a drone are as follows. Different components of the drone are shown in Fig. 3.1.

- *Flight controller*: The flight controller is simply a circuit board with attached chips. This acts as the brain of the drone and monitors and controls the action of the drone. Three main operations done by flight controller include perception, control, and communication. As the flight controllers are connected to a set of sensors, different parameters such as height, orientation, and speed are captured by the flight controller. The flight controller gathers these sensor data and sets the desired speed for each motor. As autopilot programs are in action, flight controller needs to communicate to other computer systems about its final destination.
- *Motor/speed controllers*: Brushless direct current (BLDC) motors in drones are responsible for spinning the propellers. Therefore, getting the right combination of motor propellers is very important for a drone. The electronic speed controller (ESC) receives input from the flight controller and translates the desired speed into a signal that can be understood by the motors. The motor generates the thrust to overcome the weight of the drone and achieves a lift-off.
- *Communication system*: The communication system is one important component responsible for data flow between the ground control station (GCS) and the drones. The UAV communication systems help the drones and operators to obtain the required results. The strength of the communication type affects how far the drone can fly from the controller. Radiofrequency communications are the most effective solutions for reliable communication of drones.
- *Sensors and camera*: Sensors and cameras are two important components integrated into drones. Drones are more than just a tool for aerial photography, and in recent years, it is mounted with LiDAR, thermal, and various sensors to utilize the capabilities of drones across different applications. Some of the most common sensors that are attached to drones include gyroscopes for angular velocity and tilt, accelerometers for linear movement, and magnetometers for indicating the

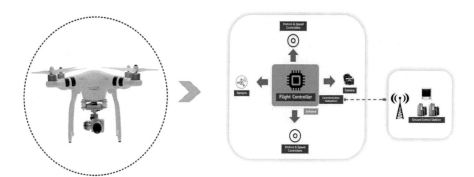

Fig. 3.1 Drone and its components

direction of a magnetic field to verify the heading. GPS determines the specific geographic locations and is often helpful to return back to the original position.

- **Drone airframe**: Drone airframe is the main skeleton that provides the support for mounting different components. Based on the objectives, the size of the frame is decided. The frame is designed to be light in weight and should have enough strength to carry the components. Materials such as plastic, fiber, aluminum, and wood are generally used for fabricating the airframes.

Optical Sensors for UAV

The primary objective of UAVs in any application is to collect high-quality images. This mandates mounting of image sensors on the UAVs. The most common sensors are in visible scale range with high resolution. But some of the applications require more detailed and hidden information even beyond the visible range of electromagnetic spectrum such as infrared, near-infrared.

- **Visible light sensors (RGB color camera)**: An RGB color camera collects images in visible light (400–700 nm) and converts them into an electrical signal and finally renders images and videos. Visible light sensors create images that replicate human vision, and images are captured with red, green, and blue channels of visible light. As visible cameras require light, their performance is also greatly affected by atmospheric conditions such as smoke, haze, fog.
- **Multispectral sensors**: While a standard visual sensor captures red, green, and blue wavelengths of light, the multispectral sensors collect the wavelengths beyond the visible spectrum, including near-infrared (NIR) and short-wave infrared (SWIR) ranges, i.e., in several broad wavelength bands. Generally, a multispectral sensor has 3–12 broad spectral bands, and it provides a broader range of options compared to RGB sensors for differentiating spectral signatures [23].
- **Hyperspectral sensors**: Hyperspectral imaging sensors are effective in capturing fine details in spectral and spatial ranges and are highly important in characterizing various properties of the specimen [2]. The spectral resolution is the key factor that differentiates hyperspectral imaging from multispectral. In hyperspectral imagery, more narrow bands are captured than multispectral in the same portion of the electromagnetic spectrum. These sensors are generally more expensive, and complex data analysis is involved in processing the data captured.
- **Thermal sensors**: Thermal sensors mounted on an UAV can measure the relative surface temperature of objects. These sensors capture infrared light in the wavelength range of 7.5–13 μm and have relatively low resolution compared to the visible sensors [46]. When the long-wave infrared light emitted from the objects hits the sensor, it heats the microbolometer resulting in a change in electrical resistance. It is then converted into electrical signals and stored as raw data. The camera forms image, differentiating the temperature variations using different color palettes.

- *LiDAR*: Light detection and ranging (LiDAR) is an active sensor that uses light energy from the laser to measure the variable distances. The time taken to reflect off an object and come back to the sensor is measured to produce high-resolution maps and 3D models of objects. LiDAR is highly accurate in measuring the density of vegetation disaster assessments, location-based investigations, etc. [38, 47].

Applications of Drones in Agriculture and Allied Sectors

Application of Drones in Agriculture

The agricultural sector is under pressure to adopt new technology and processes to maximize agricultural output, driven by a global need to feed an expanding population on a limited cultivable area [24]. However, drones should be used to their full potential in precision farming to meet end-user requirements.

Sensor data collected through drones are used with real-time data analytics in precision farming to discover spatial variability in the field to boost farm output. Drone missions produce a large amount of raw data that can be utilized to activate agricultural analytical systems. Drones can do soil health scans, monitor crop health, assist in irrigation schedule planning, apply fertilizers, estimate production statistics, and provide essential data for weather analysis as part of precision farming. When drone data are combined with data from other sources and analytic tools, actionable information is produced. Liquid insecticides, fertilizers, and herbicides are regularly applied by drones at precise variable rates. The applications of drones in agriculture are briefly explained in Table 3.1 [33].

Precision Agriculture

Drones have not yet made their way into the mainstream of precision agriculture (PA) techniques, but they are becoming increasingly important in precision agriculture, supporting agriculture specialists in leading sustainable farming practices while also protecting and boosting revenues. Drones can assist PA by performing variety of agricultural tasks including soil health scanning, seed planting, fertilizer application, crop stress management, irrigation schedule planning, weed management, crop yield management and weather analysis. Drones equipped with infrared, multispectral, and hyperspectral sensors can precisely and accurately assess crop health and soil conditions. In agricultural mapping tools, NDVI data can be combined with other indices like the crop water stress index (CWSI) and the canopy chlorophyll content index (CCCI) to provide useful insights into crop health. Drones provide a number of advantages over other aerial-based remote sensing platforms, including the following:

- Possibility of combining 3D canopy height and orthophoto data
- Rapid data collection with excellent quality, range, and resolution

Table 3.1 Applications of drones in agriculture

Objective	Example	Method	References
Crop stress management	Nitrogen stress detection	UAV RGB imagery using deep learning, U-Net model	Zhang et al. [54]
	Water status of wheat genotypes	UAV-thermal imaging and machine learning	Das et al. [12]
	Crop drought mapping system	Machine learning, UAV remote sensing RGB imagery	Su et al. [42]
	Estimation of crop emergence in potatoes	Semi-automated image analysis software and high-resolution RGB orthoimages of UAV	Li et al. [28]
	Land-surface temperature variability and orchards crop stress monitoring	Spatially coincident hyperspectral and thermal imagery	Shivers et al. [39]
	Canopy temperature as an indicator of tree stress	UAV-borne thermal systems for detecting disease and LiDAR for changes in the canopy structure	Awais et al. [6], Smigaj et al. [40]
	Crop scouting technique for gummy stem blight in watermelon	UAV-based multispectral imaging and NDVI	Kalischuk et al. [27]
Weed management	Mapping invasive plant	UAV-derived 2D digital orthomap and 3D mesh model	Hunter et al. [22], Wu et al. [51]
	Weed discrimination in maize	UAV-based hyperspectral imaging	Casa et al. [10]
	Automatic UAV-based detection in vineyards	Object-based image analysis (OBIA) on UAV and multispectral cameras (RGB and RGNIR)	Jimenez-Brenes et al. [25]
	Automated weed detection in corn and soybean fields	Machine learning approaches and UAV-based sensors like multispectral and thermal imagery	Etienne and Saraswat [13]
Crop-soil spatial and temporal variability	Quantitative estimation of soil salinity	UAV-borne hyperspectral and satellite multispectral images	Falco et al. [14], Hu et al. [21]
	Vineyard variability assessment	Comparison of satellite and UAV-based multispectral imagery	Campos et al. [8]
	Monitoring drought effects on vegetation productivity	Satellite solar-induced chlorophyll fluorescence	Gomez et al. [17]
	Free water table area monitoring	Satellite and UAV orthophoto maps	Goraj et al. [19]
Application of pesticides and fertilizers	Variable-rate spraying technology	UAV, artificial neural network (ANN) technology, an error back propagation (BP), and neural network model	Wang et al. [48], Wen et al. [50]

(continued)

Table 3.1 (continued)

Objective	Example	Method	References
	Compound control method for pesticide spraying	UAV, Lyapunov theory, radial basis function neural network (RBFNN), weight matrices of the RBFNN	Liao et al. [29]
	Effect of spray volume on deposition and the control of pests and disease	UAV, three-level spray volumes, three nozzle sizes, electric air-pressure knapsack sprayer	Wang et al. [49]
Irrigation management	Estimation of soil moisture	Machine learning techniques and unmanned aerial vehicle (UAV) multispectral imagery	Aboutalebi et al. [1]
	Precision irrigation management	UAV, evapotranspiration (ET) coefficient, RGB, and near-infrared (NIR) sensors	Chen et al. [11]; Ronchetti et al. [37]
	Irrigation irregularities in an olive grove	UAV, multispectral camera, NDVI, GNDVI, SAVI, and NDRE vegetation indices	Jorge et al. [26], Yang et al. [52]
Crop yield management	Predicting grain yield in rice	Multitemporal vegetation indices from UAV-based multispectral and digital imagery	Zhou et al. [56]
	Yield estimation in cotton	UAV, multiple image sensors, digital surface model (DSM), triangular greenness index (TGI), NDVI	Feng et al. [15]
	Rice grain yield estimation at the ripening stage	UAV-based remotely sensed images, convolution neural network (CNN), vegetation index (VI)	Furukawa et al. [16]
	Estimate of winter-wheat yield	UAV, ultrahigh-ground-resolution image textures and vegetation indices	Ramos et al. [36]
	Modeling maize aboveground	Machine learning regression algorithms, support vector machine, artificial neural network, and random forest	Han et al. [20]
	Estimation of rice aboveground biomass	UAV, multispectral (MS) camera, normalized difference texture index (NDTI)	Zheng et al. [55]

- Capacity to acquire multiangular data (particularly from snapshot cameras)
- Capacity to run many sensors at the same time.

Weed and Pest Management

Weed mapping is one of the most prominent uses of drones in precision agriculture. Weeds can cause complications during harvesting, in addition to causing problems with crop growth. To solve the aforementioned concerns, drones can be used for site-specific weed management (SSWM). Drones have also been used to spray herbicides spatially based on weed density assessed by drone-collected hyperspectral images. Because weed plants usually grow in a few regions of the field, the field is divided into management zones, each of which receives its own treatment.

Spraying drones can help reduce operator exposure while also improving the capacity to distribute chemicals in a timely and spatially resolved manner. With the use of precision distance measurement technology, drones can follow the curve of the ground while maintaining a constant height. As a result, the drone may spray the appropriate amount of herbicide based on the crop site, changing both its height and the amount of herbicide sprayed. Finally, drone-based technologies have the potential to improve agricultural spray management significantly.

Irrigation and Nutrient Management

Crop irrigation consumes 70% of the world's water, emphasizing the significance of precision irrigation. Farmers can save time and water by recognizing areas that need a lot of water. At the same time, precision farming techniques can increase crop yield and quality. The field is divided into multiple irrigation zones in precision agriculture to manage the resources precisely. Drone technology can be utilized in precision agriculture to manage crop irrigation. Drones can address water body characterization, crop water requirements, and crop water stress monitoring by utilizing new methodologies/approaches, such as satellites and the most advanced multispectral/thermal cameras attached in drones with high flight capacity/autonomy (Fig. 3.2). Using drones fitted with the necessary sensors, it is possible to identify areas of a crop with water and nutrient stress that demand more fertigation. Simultaneously, these technologies enable the creation of specific maps that depict the real-time soil moisture content for efficient planning of irrigation.

The land-surface temperature may be used to compute the crop water stress index (CWSI) through the empirical model presented in Eq. (3.1).

$$\text{CWSI} = (T_c - T_{\text{wet}})/(T_{\text{dry}} - T_{\text{wet}}) \tag{3.1}$$

where T_c = crop canopy temperature, and T_{wet} and T_{dry} are the lower and upper boundary temperatures, corresponding to a fully transpiring leaf with open stomata and a non-transpiring leaf with closed stomata, respectively. These figures can be obtained in the field by using thermal infrared imagery captured by drones [30, 34]. CWSI values range from 0 (no stress) to 1 (severe stress).

Fig. 3.2 Drone with
multispectral camera for
irrigation scheduling

Crop and Soil Monitoring

Drones that can fly at low altitudes may be utilized in field mapping using various remote sensing approaches. Drone equipped with remote sensing equipment allows growers to gather, visualize, and analyze crop and soil health conditions at various stages of production in a simple and cost-effective manner. Crop biomass and nitrogen status, as well as yield estimation, have been the focus of several recent studies. The most common crop attribute is biomass, which can be used to determine if additional fertilizer or other treatments are necessary when combined with information on nitrogen content. The data obtained by the drones can also be utilized to construct field maps and measure other characteristics such as crop height, distance between rows or between plants, and leaf area index (LAI). The normalized difference vegetation index (NDVI) deserves special attention since it is widely used in agricultural applications to estimate a variety of crop-related metrics, including biomass, canopy structure, leaf area index (LAI), crop management, and crop zone mapping [9, 34]. Drones fitted with various controllers and sensors such as control systems and radio remote control can perform multiple tasks and can assist in real-time mapping of agricultural fields. Aerial-based remote sensing platforms like drones are fitted with multispectral cameras, digital cameras, infrared thermal imagers, hyperspectral sensors, and light detection and ranging (LIDAR) sensors. Studies have established that aerial remote sensing using drone can acquire plant canopy temperature and NDVI of breeding plots effectively. The multispectral camera takes images and stores them in memory and sends them to the ground station through telemetry. The geographic indicator, viz. normalized difference vegetation index (NDVI), represented in Eq. 3.2, was used to examine data from the multispectral camera through telemetry [7, 32].

$$\text{NDVI} = (R_{NIR} - R_{RED})/(R_{NIR} + R_{RED}) \qquad (3.2)$$

where R_{NIR} = Reflectance of the near infrared band and R_{RED} = Reflectance of the red band.

The results range from -1 to $+1$, with a value close to 0 (zero) indicating no vegetation and a value close to $+1$ (0.8–0.9) indicating the highest density of green leaves. Farmers can easily identify the field where pesticides can be sprayed based on these findings.

Monitoring and analysis of soil conditions are crucial for modern farming practices. Drones can make the soil sampling proactive and help farmers foresee potential issues. In a few number of flights, drones can conduct soil analysis in real time using multispectral cameras. From the insights collected using the drone-based imaging, farmers and other stakeholders can determine the ideal seeding pattern, effective watering and fertilizer strategies, etc. Multispectral imaging can produce quality soil maps and capture the data on regular basis to help the farmers to understand how the crop production is impacting the soil health over a longer period.

Survey of Farm Lands and Land-Use Mapping

A drone can inspect a bigger area of agricultural land with much higher spatial and temporal details. Drones have been used in agriculture to produce high-resolution photographs that can distinguish individual crops and weeds on a miniature scale. Drones fitted with LIDAR sensors can be used to survey and monitor orchard fields. Surveying the nutritional condition of soil at different soil types, nutrient ranges, and nutrient requirements within and between fields is another area where drones can be used. Drones can be used to reduce stubble burning in agricultural fields in India, where it is a big issue and a threat to the environment. Pests, insects, locusts, and armyworms can be controlled using multispectral imagery acquired through drones.

Flood and Environment Risk Assessment

Drone platforms have recently seen a rapid surge in their use for environmental monitoring. Drones are largely employed in hydrological research to collect precise and current geographical data on the river environment, as well as to monitor fluvial processes with high precision and sample frequency. They provide low-cost and high-resolution environmental data. UAVs stand out among the applications dedicated to the study of floods because of the times in which they can be utilized, i.e., before (prevention), during, and after occurrence. They have many uses for wetland mapping and hydrological modeling (e.g., damage assessment, remapping of the affected area). UAV applications help in planning and preparation of emergency response during flood as well as the development of tools to help with reaction before, during, and after the disaster [41]. Drone technology is rapidly advancing, allowing for quantitative estimation of hydraulic data such as inundated zones and

Fig. 3.3 Use of drone for orchards' yield estimation

surface flow measurements. Drone-borne observations have also been used to forecast hydrological variables like evaporation and evapotranspiration using energy balance models using soil and vegetation data input. Drone-borne images of geomorphologic variables and sediment suspension have also been investigated, as well as the velocity of surface water in a river.

Horticulture

Horticulture is a growing industry for drones as it is helping the growers stay ahead of major issues before they are becoming too complex to solve. Drones can provide numerous solutions to horticulture domain like crop characterization and counting, crop growth analysis, yield estimation, etc. (Fig. 3.3). Some of the key applications of drones in horticulture are as follows:

- Drone imaging coupled with advanced computer vision algorithms can deliver automated plant counts, plant health assessments, and plant density.
- Conventional yield monitoring practices can be replaced by drone-based crop yield monitoring and scouting.
- Drones are also being used for increasing water-use efficiency in orchards and detecting possible leaks in irrigation system.

Forestry

Drones have been utilized in forestry to create an integrated information system for forest conservation. Massive and high-resolution orthomaps are created by stitching together drone pictures. These orthomaps can be used in GIS systems to analyze, plan, and manage data. Drone technology is being used to improve forest management and operations planning, as well as to monitor illegal activity and encroachment. It also assists in collection of forest metrics such as carbon sequestration, tree canopy analysis, conservation characteristics, native species tracking, biodiversity monitoring, and ecological landscape aspects, among others. Because of the high resolution of the sensors, UAVs can also detect wildlife damage. When looking for wildlife damage, more information about the animals' habits is required, or experts' field investigations are required to avoid false interpretation of the missing vegetation [31]. The important applications of UAVs for forestry purposes are as follows [44]:

- Estimation of dendrometric parameter
- Classification of tree species
- Measurement of forest spatial gaps
- Forest fire evaluation and post-fire restoration monitoring
- Forest disease tracing and forest health surveillance
- Estimation of post-harvest soil displacement.

Fisheries

Although the Food and Agriculture Organization (FAO) has assessed the continual decline of marine fish resources, many interventions have been made by government institutions, corporate groups, and people to raise awareness of the global fishery resources [5, 45]. Fishing is one of the most important food-producing enterprises. As the size of the fishing fleet has risen over time, so has the demand for sophisticated electronic equipment and components including sensors, sounders, and sonars. The drone can be used to detect fish by providing a bird's eye view of the surrounding water surface beyond what normal human vision can see. It aids in detection of fish in the presence of predatory birds, shadows, and other factors. This minimizes ideal waiting time, saves fuel, lowers costs, enhances productivity, and minimizes illicit fishing practices such as the employment of pair trawlers and purse seiners, which are prohibited in India because they harm the ecosystem [3]. Traditional fishery water assessment and prediction rely on collecting water samples and sending them to inspection over a period of time, followed by physical and chemical analysis and testing. This takes a long time and consumes a lot of human resources, which is not only inefficient but also inconvenient. A method for measuring and predicting fishery water quality employing drone-based electrochemical sensor arrays such as pH, dissolved oxygen, and ammonia nitrogen was presented [53]. The important applications of drones for aquaculture are as follows [45]:

- Fish site surveillance and mapping
- Fish farm monitoring and management
- Fish feeding management
- Fish behavior monitoring
- Water quality monitoring and assessment.

Livestock Management

One of the most promising applications for drones is livestock husbandry, where UAVs facilitate various operations for efficient animal management. However, there are numerous environmental, technical, economic, and strategic challenges in the domain. The use of advanced technological techniques such as artificial intelligence (AI), the Internet of Things (IoT), machine learning (ML), deep learning (DL), advanced sensors, and other advanced sensors, as well as assurances of animal welfare while operating drones, could lead to widespread adoption of drone technology among livestock farmers. Detection, counting the numbers, identifying the types, tracking while grazing, monitoring health issues, rounding up the cattle, behavior monitoring, and calculating the herd distribution are parts of livestock monitoring that drones can do. Drones have been in great demand in livestock management for sanitization of farm areas as well as inside the animal sheds by spraying the sanitizers, conducting behavior studies/phenomics and onsite delivery of semen/vaccines/medicaments/fertile eggs. Drones can also be employed for reaching far flung areas for delivering vital inputs as well as data collection and recording of livestock and poultry management practices. Additionally, it can be utilized for animal/poultry population enumerations particularly for the nomadic/pastoralist populations, tracking the home tract as well as migration of livestock population and mapping feed and fodder grasses' areas.

Conclusions

Traditional agricultural practices are giving way to a new "intelligent" approach to the farming process. The use of drone, sensors and control system, can lead to a novel precision farming approach that maximizes the potential of agricultural crops. Drones can offer various solutions in many areas of agriculture and allied sciences such as yield estimation, spraying, scouting and bird scaring, crop health monitoring, forest mapping, fish farm monitoring, livestock tracking. These key emerging technologies have the potential to increase crop yield and quality, lower costs, and reduce the environmental impact of traditional farming. Drone-based agriculture has a wide range of applications on temporal and spatial scales, making it an effective modernistic technology to address the complex problem of food insecurity.

References

1. Aboutalebi, M., Allen, L.N., Torres-Rua, A.F., McKee, M., Coopmans, C.: Estimation of soil moisture at different soil levels using machine learning techniques and unmanned aerial vehicle (UAV) multispectral imagery. In: Autonomous Air and Ground Sensing Systems for Agricultural Optimization and Phenotyping, vol. IV, no. 11008, pp. 216–226 (2019)
2. Adão, T., Hruška, J., Pádua, L., Bessa, J., Peres, E., Morais, R., & Sousa, J. J. Hyperspectral imaging: A review on UAV-based sensors, data processing and applications for agriculture and forestry. Remote sens. 9(11), 1110 (2017)

3. Ahilan, T., Adityan, V.A., Kailash, S.: Efficient utilization of unmanned aerial vehicle (UAV) for fishing through surveillance for fishermen. Int. J. Aerosp. Mech. Eng. **9**(8), 1468–1471 (2015)
4. Anonymous: The Drone Rules 2021. Ministry of Civil Aviation, Government of India (2022)
5. Anonymous: FAO. The State of World Fisheries and Aquaculture 2020. Sustainability in Action, FAO, Rome, Italy (2020)
6. Awais, M., Li, W., Cheema, M.J.M., Hussain, S., AlGarni, T.S., Liu, C., Ali, A.: Remotely sensed identification of canopy characteristics using UAV-based imagery under unstable environmental conditions. Environ. Technol. Innov. **22**, 101465 (2021)
7. Bhandari, A.K., Kumar, A., Singh, G.K.: Feature extraction using normalized difference vegetation index (NDVI): a case study of Jabalpur city. Procedia Technol. **6**, 612–621 (2012)
8. Campos, J., García-Ruíz, F., Gil, E.: Assessment of vineyard canopy characteristics from vigour maps obtained using UAV and satellite imagery. Sensors **21**(7), 2363 (2021)
9. Candiago, S., Remondino, F., De Giglio, M., Dubbini, M., Gattelli, M.: Evaluating multispectral images and vegetation indices for precision farming applications from UAV images. Remote Sensing **7**(4), 4026–4047 (2015)
10. Casa, R., Pascucci, S., Pignatti, S., Palombo, A., Nanni, U., Harfouche, A., Laura, L., Di Rocco, M., Fantozzi, P.: UAV-based hyperspectral imaging for weed discrimination in maize. Precision Agric. **19**, 24–35 (2019)
11. Chen, A., Orlov-Levin, V., Meron, M.: Applying high-resolution visible-channel aerial imaging of crop canopy to precision irrigation management. Agric. Water Manag. **216**, 196–205 (2019)
12. Das, S., Christopher, J., Apan, A., Choudhury, M.R., Chapman, S., Menzies, N.W., Dang, Y.P.: Evaluation of water status of wheat genotypes to aid prediction of yield on sodic soils using UAV-thermal imaging and machine learning. Agric. For. Meteorol. **307**, 108477 (2021)
13. Etienne, A., Saraswat, D.: Machine learning approaches to automate weed detection by UAV based sensors. In: Autonomous Air and Ground Sensing Systems for Agricultural Optimization and Phenotyping, vol. IV, no. 11008, pp. 202–215 (2019)
14. Falco, N., Wainwright, H.M., Dafflon, B., Ulrich, C., Soom, F., Peterson, J.E., Brown, J.B., Schaettle, K.B., Williamson, M., Cothren, J.D., Ham, R.G.: Influence of soil heterogeneity on soybean plant development and crop yield evaluated using time-series of UAV and ground-based geophysical imagery. Sci. Rep. **11**(1), 1–17 (2021)
15. Feng, A., Zhou, J., Vories, E.D., Sudduth, K.A., Zhang, M.: Yield estimation in cotton using UAV-based multi-sensor imagery. Biosys. Eng. **193**, 101–114 (2020)
16. Furukawa, F., Maruyama, K., Saito, Y.K., Kaneko, M.: Corn height estimation using UAV for yield prediction and crop monitoring. In: Unmanned Aerial Vehicle: Applications in Agriculture and Environment, pp. 51–69. Springer, Cham (2020)
17. Gómez-Gálvez, F.J., Pérez-Mohedano, D., de la Rosa-Navarro, R., Belaj, A.: High-throughput analysis of the canopy traits in the worldwide olive germplasm bank of Córdoba using very high-resolution imagery acquired from unmanned aerial vehicle (UAV). Sci. Hortic. **278**, 109851 (2021)
18. González-Jorge, H., Martínez-Sánchez, J., Bueno, M., Arias, A.P.: Unmanned aerial systems for civil applications: a review. Drones **1**, 2 (2017)
19. Góraj, M., Wróblewski, C., Ciężkowski, W., Jóźwiak, J., Chormański, J.: Free water table area monitoring on wetlands using satellite and UAV orthophotomaps-Kampinos National Park case study. Meteorol. Hydrol. Water Manage. Res. Oper. Appl. 7 (2019)
20. Han, L., Yang, G., Dai, H., Xu, B., Yang, H., Feng, H., Li, Z., Yang, X.: Modeling maize above-ground biomass based on machine learning approaches using UAV remote-sensing data. Plant Methods **15**(1), 1–19 (2019)
21. Hu, B., Zhou, Y., Jiang, Y., Ji, W., Fu, Z., Shao, S., Li, S., Huang, M., Zhou, L., Shi, Z.: Spatio-temporal variation and source changes of potentially toxic elements in soil on a typical plain of the Yangtze River Delta, China. J. Environ. Manage. **271**, 110943 (2020)
22. Hunter, J.E., III., Gannon, T.W., Richardson, R.J., Yelverton, F.H., Leon, R.G.: Integration of remote-weed mapping and an autonomous spraying unmanned aerial vehicle for site-specific weed management. Pest Manag. Sci. **76**(4), 1386–1392 (2020)

23. Iost Filho, F.H., Heldens, W.B., Kong, Z., de Lange, E.S.: Drones: innovative technology for use in precision pest management. J. Econ. Entomol. **113**, 1–25 (2020)
24. Jarman, M., Vesey, J., Febvre, P.: Unmanned Aerial Vehicles (UAVs) for UK Agriculture: Creating an Invisible Precision Farming Technology. White Paper (2016)
25. Jiménez-Brenes, F.M., Lopez-Granados, F., Torres-Sánchez, J., Peña, J.M., Ramírez, P., Castillejo-González, I.L., de Castro, A.I.: Automatic UAV-based detection of Cynodon dactylon for site-specific vineyard management. PLoS ONE **14**(6), 0218132 (2019)
26. Jorge, J., Vallbé, M., Soler, J.A.: Detection of irrigation inhomogeneities in an olive grove using the NDRE vegetation index obtained from UAV images. Eur. J. Remote Sens. **52**(1), 169–177 (2019)
27. Kalischuk, M., Paret, M.L., Freeman, J.H., Raj, D., da Silva, S., Eubanks, S., Wiggins, Z., Lollar, M., Marois, J.J., Mellinger, H.C., Das, J.: An improved crop scouting technique incorporating UAV-assisted multispectral crop imaging into conventional scouting practice for gummy stem blight in watermelon. Plant Dis. First Look (2019)
28. Li, B., Xu, X., Han, J., Zhang, L., Bian, C., Jin, L., Liu, J.: The estimation of crop emergence in potatoes by UAV RGB imagery. Plant Methods **15**(1), 1–13 (2019)
29. Liao, S., Lei, X., Xiao, Y.: The compound control method for pesticide spraying quadrotor UAVs. In: IEEE 3rd Information Technology, Networking, Electronic and Automation Control Conference, pp.1022–1027. IEEE (2019)
30. Matese, A., Baraldi, R., Berton, A., Cesaraccio, C., Di Gennaro, S.F., Duce, P., Facini, O., Mameli, M.G., Piga, A., Zaldei, A.: Estimation of water stress in grapevines using proximal and remote sensing methods. Remote Sens. **10**, 114 (2018)
31. Milics, G.: Application of UAVs in precision agriculture. In: International Climate Protection, pp. 93–97. Springer (2019)
32. Mogili, U.R., Deepak, B.B.V.L.: Review on application of drone systems in precision agriculture. Procedia Comput. Sci. **133**, 502–509 (2018)
33. Mukherjee, A., Misra, S., Raghuwanshi, N.S.: A survey of unmanned aerial sensing solutions in precision agriculture. J. Netw. Comput. Appl. **148**, 102461 (2019)
34. Padua, L., Marques, P., Adão, T., Guimarães, N., Sousa, A., Peres, E., Sousa, J.J.: Vineyard variability analysis through UAV-based vigour maps to assess climate change impacts. Agronomy **9**(10), 581 (2019)
35. Radoglou-Grammatikis, P., Sarigiannidis, P., Lagkas, T., Moscholios, I.: A compilation of UAV applications for precision agriculture. Comput. Netw. **172**, 107148 (2020)
36. Ramos, A.P.M., Osco, L.P., Furuya, D.E.G., Gonçalves, W.N., Santana, D.C., Teodoro, L.P.R., da Silva Junior, C.A., Capristo-Silva, G.F., Li, J., Baio, F.H.R., Junior, J.M.: A random forest ranking approach to predict yield in maize with uav-based vegetation spectral indices. Comput. Electron. Agric. **178**, 105791 (2020)
37. Ronchetti, G., Mayer, A., Facchi, A., Ortuani, B., Sona, G.: Crop row detection through UAV surveys to optimize on-farm irrigation management. Remote Sens. **12**(12), 1967 (2020)
38. Sankey, T., Donager, J., McVay, J., Sankey, J.B.: UAV LIDAR and hyperspectral fusion for forest monitoring in the southwestern USA. Remote Sens. Environ. **195**, 30–43 (2017)
39. Shivers, S.W., Roberts, D.A., McFadden, J.P.: Using paired thermal and hyperspectral aerial imagery to quantify land surface temperature variability and assess crop stress within California orchards. Remote Sens. Environ. **222**, 215–231 (2019)
40. Smigaj, M., Gaulton, R., Suárez, J.C., Barr, S.L.: Canopy temperature from an unmanned aerial vehicle as an indicator of tree stress associated with red band needle blight severity. For. Ecol. Manage. **433**, 699–708 (2019)
41. Stephan, F., Reinsperger, N., Grünthal, M., Paulicke, D., Jahn, P.: Human drone interaction in delivery of medical supplies: a scoping review of experimental studies. PLoS ONE **17**(4), 0267664 (2022)
42. Su, J., Coombes, M., Liu, C., Zhu, Y., Song, X., Fang, S., Guo, L., Chen, W.H.: Machine learning-based crop drought mapping system by UAV remote sensing RGB imagery. Un. Syst. **8**(1), 71–83 (2020)

43. Toriyama, K.: Development of precision agriculture and ICT application thereof to manage spatial variability of crop growth. Soil Sci. Plant Nutr. **66**, 811–819 (2020)
44. Torresan, C., Berton, A., Carotenuto, F., Di Gennaro, S.F., Gioli, B., Matese, A., Miglietta, F., Vagnoli, C., Zaldei, A., Wallace, L.: Forestry applications of UAVs in Europe: a review. Int. J. Remote Sens. **38**(8–10), 2427–2447 (2017)
45. Ubina, N.A., Cheng, S.C.: A review of unmanned system technologies with its application to aquaculture farm monitoring and management. Drones **6**(1), 12 (2022)
46. Van der Merwe, D., Burchfield, D.R., Witt, T.D., Price, K.P., Sharda, A.: Drones in agriculture. In: Sparks, D.L. (ed.) Advances in Agronomy, pp. 1–30. Academic Press (2020)
47. Wallace, L., Lucieer, A., Watson, C., Turner, D.: Development of a UAV-LiDAR system with application to forest inventory. Remote Sens. **4**, 1519–1543 (2012)
48. Wang, G., Lan, Y., Qi, H., Chen, P., Hewitt, A., Han, Y.: Field evaluation of an unmanned aerial vehicle (UAV) sprayer: effect of spray volume on deposition and the control of pests and disease in wheat. Pest Manag. Sci. **75**(6), 1546–1555 (2019)
49. Wang, L., Lan, Y., Yue, X., Ling, K., Cen, Z., Cheng, Z., Liu, Y., Wang, J.: Vision-based adaptive variable rate spraying approach for unmanned aerial vehicles. Int. J. Agric. Biol. Eng. **12**(3), 18–26 (2019)
50. Wen, S., Zhang, Q., Yin, X., Lan, Y., Zhang, J., Ge, Y.: Design of plant protection UAV variable spray system based on neural networks. Sensors **19**(5), 1112 (2019)
51. Wu, Z., Ni, M., Hu, Z., Wang, J., Li, Q., Wu, G.: Mapping invasive plant with UAV-derived 3D mesh model in mountain area—a case study in Shenzhen Coast, China. Int. J. Appl. Earth Obs. Geoinf. **77**, 129–139 (2019)
52. Yang, S., Yu, W., Yang, L., Du, B., Chen, S., Sun, W., Jiang, H., Xie, M., Tang, J.: Occurrence and fate of steroid estrogens in a Chinese typical concentrated dairy farm and slurry irrigated soil. J. Agric. Food Chem. **69**(1), 67–77 (2020)
53. Yao, D., Cheng, L., Wu, Q., Zhang, G., Wu, B., He, Y.: Assessment and prediction of fishery water quality using electrochemical sensor array carried by UAV. In: 2019 IEEE International Symposium on Olfaction and Electronic Nose (ISOEN), pp.1–4 (2019)
54. Zhang, J., Xie, T., Yang, C., Song, H., Jiang, Z., Zhou, G., Zhang, D., Feng, H., Xie, J.: Segmenting purple rapeseed leaves in the field from UAV RGB imagery using deep learning as an auxiliary means for nitrogen stress detection. Remote Sens. **12**(9), 1403 (2020)
55. Zheng, H., Cheng, T., Zhou, M., Li, D., Yao, X., Tian, Y., Cao, W., Zhu, Y.: Improved estimation of rice aboveground biomass combining textural and spectral analysis of UAV imagery. Precision Agric. **20**(3), 611–629 (2019)
56. Zhou, X., Zheng, H.B., Xu, X.Q., He, J.Y., Ge, X.K., Yao, X., Cheng, T., Zhu, Y., Cao, W.X., Tian, Y.C.: Predicting grain yield in rice using multi-temporal vegetation indices from UAV-based multispectral and digital imagery. ISPRS J. Photogramm. Remote. Sens. **130**, 246–255 (2017)

Chapter 4
Achieving Agriculture 4.0 Through Modernization and Enhancement with Mechanization, Automation and Advanced Technologies

Seng Teik Ten, Khairul Anuar Bin Shafie, and Rohazrin Bin Abdul Rani

Introduction

Food security, according to the United Nations, is defined as the availability of and adequate access to sufficient, safe and nutritious food at all times in order to live a healthy and active life [1]. In order to fulfil the global food security, food production must be considerably increased to maintain the world population, which is expected to reach 9.7 billion people by 2050 [2]. Recently, humans are confronted with huge challenges in agricultural sectors with the issues such as diminishing natural resources and lands, global warming and unpredictable weather patterns. Currently, the impact of the COVID-19 pandemic is making the situation even worst on agricultural food productivity, which has made food security a major concern around the world. Overall, the COVID-19 pandemic during the first quarter of 2020 is expected to result in a 3.11% or 17.03 million tons reduction in Southeast Asia's aggregate volume of agricultural production due to a decline in agricultural farm labor affecting 100.77 million people [3]. During the COVID-19 pandemic, Malaysia imposed a Movement Control Order (MCO) to prevent the pandemic from spreading. All economic, social, agricultural and other activities were completely halted. Malaysia's agriculture supply chain has been disrupted as a result of the action. As a result, agricultural food industries have to be modernized and enhanced in order to significantly increase food productivity; otherwise, achieving the United

S. T. Ten (✉) · K. A. B. Shafie · R. B. Abdul Rani
Engineering Research Centre, Malaysian Agricultural Research and Development Institute (MARDI), Serdang, Malaysia
e-mail: stten@mardi.gov.my

K. A. B. Shafie
e-mail: khanuar@mardi.gov.my

R. B. Abdul Rani
e-mail: rohazrin@mardi.gov.my

© Centre for Science and Technology of the Non-aligned and Other Developing Countries 2023
K. Pakeerathan (ed.), *Smart Agriculture for Developing Nations*, Advanced Technologies and Societal Change, https://doi.org/10.1007/978-981-19-8738-0_4

Nations Sustainable Development Goals, Zero Hunger, will be difficult. On the bright side, in this Industry Revolution 4.0 era, technologies are able to be adopted in agricultural technologies to modernize and enhance the agricultural industries. Thus, this paper aims to review the existing agricultural technologies which are able to be transformed from mechanization, automation and toward smart agriculture, especially the research works from MARDI.

Mechanization in Agricultural Production

Agricultural tools, equipment and machinery for land preparation, planting, crop management, harvest and post-harvest activities, processing and all actions in the agri-food value chain are examples of mechanization options. The mechanized land preparation implements are quite general for field crops which are plough, harrow and rotovator. However, the paddy field land preparation needs additional land leveling implement. Mechanized land preparation is a well-established technique, particularly on mineral soils, and it helps to boost farm productivity. On problematic soil, such as peat, a special prime mover is required as a mechanization enabler. To overcome low bearing capacity soil, a tractor equipped with a rubber track system is used (Fig. 4.1). The mechanized planting machinery is able to reduce labor force and activity time; for example, the mechanical pineapple transplanter (Fig. 4.2), which was developed by MARDI, transplants a hectare in 20 h, compared to 80 h for manual transplanting. The mechanized crop management package can be designed for multiple functions in one operation. It consists of weeding and chemical inputs' application; for example, the inter-row cultivator cum fertilizer applicator can perform two functions at once, weeding and granule fertilizer application. Another example is the mechanized chemical sprayer with a 12-m boom and 800-L tank (Fig. 4.3). It is used for spraying weedicides, insecticides and even flowering hormone. The machine is up to 5 times faster than manual operation using knapsack spraying machine. A fruit harvesting machine was developed to overcome the problem of the harvesting operation which is one of the most laborious farm operations that is consuming the most man-hours per hectare. For example, in pineapple plantation, the machine features a 12-m boom with a rubber conveyor (Fig. 4.4). The fruits are manually chopped by a group of three field workers and placed on a rubber conveyor that transports the fruit to a collecting box mounted on a trailer during the harvesting operation. The harvesting operation can be completed in one hour for a one-hectare plot. Mechanized crop residue management can reduce the use of chemicals to kill unwanted crop as well as the air pollution caused by open burning. In the pineapple production system; for example, before replanting activity, a specially designed rotovator (Fig. 4.5) is used to shred and plough the plant residue into the soil. The technique is not only environmentally friendly, but it also consumes less time and labor. In the paddy production system, a collecting machine known as a baler is used to collect paddy straw and then convert it into value-added products such as compost, rope and charcoal. For the post-harvest and processing activities, machineries have been established, such

Fig. 4.1 Rubber tracked tractor

Fig. 4.2 Mechanical pineapple transplanter

as Jackfruit Bulbs Extractor which is capable to extract edible bulbs from jackfruit skin almost 3 times faster compared to manual method.

Adoption Industry 4.0 Technologies in Agricultural Production

Agriculture evolves with science and technology, and it is adopting IR4.0 technologies to become the new paradigm, Agriculture 4.0 [4–6]. Within this paradigm, land preparation, planting, crop management, including weeding and pest control, and harvesting are mainly driven by digitalization, automation, Internet of things (IoT),

Fig. 4.3 12-m boom spraying machine

Fig. 4.4 Pineapple harvesting aid

Fig. 4.5 Specially designed rotovator for pineapple crop residue management

robotics and artificial intelligence [5]. There are mainly two options to achieve Agriculture 4.0, integrate the IR4.0 technologies into the existing system or construct the system from the scratch if the existing system is not available yet or not able to be integrated [7]. For the first option, on field mechanization, the prime mover or tractor can be transformed to unmanned guided vehicle (UGV) autonomous by adding various types of sensors, such as ultrasonic, compass, gyroscope, accelerometer and artificial intelligence (AI) system in the on-board computer [8]. At the same time, with the field topography, mapping can be created by the integration of the advent of GPS and GNSS of precision agriculture research with the advanced sensor, light detection and ranging (LiDAR) system to form the smart navigation system [9]. These transformed autonomous machines can prepare the land without human intervention. It can communicate and negotiate with other tractors about what duties need to be completed, such as bed preparation and field tilling. In the land preparation stage, light autonomous soil parameter taking robot can be implemented to generate a soil fertility map. Combination of the MARDI-developed smart autonomous and the automatic index system will become the smart automatic leveling machine [10] for land preparation in paddy field (Fig. 4.6). For the crop management, fertilizer and pesticide application can be smartly done by the combination of unmanned aerial vehicles (UAVs) and autonomous fertilizer application or chemical spraying system using variable-rate technology (VRT) in rice production. A GIS-based mapping on soil fertility and plant condition obtained by UAV is processed and analyzed with AI system to produce the treatment map. The treatment map allows site-specific crop management for the smart VRT autonomous or drone system to apply fertilizer and pesticide with the right amount, at the right place and on the right time. Thus, it can optimize the usage of fertilizer and pesticide without wastage and can be done without human intervention as inhaling chemical can threaten the human's health. Beyond outdoor crop management, the indoor controlled environment can be upgraded; MARDI is also conducting the R&D in producing vegetables in the factory, plant factory (PL). The PL is with the low energy consumption vertical farming building which is fully integrated with smart IoT-based control and monitoring system. The system can be fully monitored and controlled by the AI system or overridden by remotely monitoring and controlling the microclimate of the building, including the temperature, humidity, CO_2 level, air flow, fertigation system and artificial lighting to the plants.

The controlled environment system consists of three main components (Fig. 4.7): mobile application system (e-farm application), environment sensors and repeater module. The sensors installed in the plant factory are temperature sensors, humidity sensors, CO_2 sensors and air flow sensors. The e-farm application was developed on Android operating platform. The data will be displayed on the tablet, and control can be made in the mobile application features to ensure optimum growth condition in the plant factory. The optimum temperature range for lettuce is estimated in between 23 and 28 °C, and the relative humidity is in between 60 and 70%. Monitoring and control system are essentials for micro-irrigation system, especially in plant factory to precisely meet the optimum plant growth requirement. The plant requires specific nutrient level depending on the growth condition. The micro-irrigation consists of

Fig. 4.6 Smart automatic
leveling machine for land
preparation in paddy field
[10]

fertigation tank, water pump, electrical injector for fertilizer and a control system. The sensor system consists of pH sensor, electrical conductivity (E.C) sensor and water temperature sensor. The sensors are connected to supervisory control and data acquisition (SCADA) system which is being used to set the parameter for E.C and pH, and the data can be visualized through the SCADA control panel. The irrigation data are also stored on a local server which is located in a room called controlled room. All data are synchronized to cloud storage, and the data are displayed on the dashboard. Figure 4.8 shows the micro-irrigation system for plant factory.

These systems have been integrated into IoT system (Fig. 4.9) by modifying the connectivity port and extended it to WiFi and 4G LTE connection. Both data are transmitted and received via local server before displaying its data on the dashboard (Fig. 4.10).

This PL based on smart farming is expected to be able to increase the productivity per unit area to 4–6 times vegetables which are high quality, nutritious and pesticide-free. The PL based is pesticide-free. Another similar system is an automatic IoT-controlled environment mushroom house (CEMH) which is developed, especially for the sudden weather change which is usually happening in Malaysia. This CEMH system is able to produce at least 30% more yield than ordinary mushroom houses, and the contamination rate is successfully kept below 2% and can provide the research facility for the high nutritional and medicinal value mushrooms as the micro-climate can be smartly controlled, monitored and analyzed. The combination of advanced imaging technology and artificial intelligence techniques is currently used in pest and disease detection, such as pest classification using the deep transfer learning technique (Classification of Vegetable Plant Pests Using Deep Transfer Learning) to train the pest classification model with the minimum number of samples because collecting a large number of samples is impractical and time-consuming. Utilizing a pre-trained convolutional neural network (CNN) architecture [11], the classes of vegetable pests can be learned. In addition, data augmentation and transfer learning approaches are applied to fine-tune the classification accuracy to 99.88%. On the other side, the Distance Diagnostic and Identification System (DDIS) is also utilized to rapidly identify pests and illnesses. This technology delivers a digital picture library with accompanying GPS location, crop and pest information [12]. Currently, MARDI adopts a mix of modern image technology and artificial intelligence to identify the

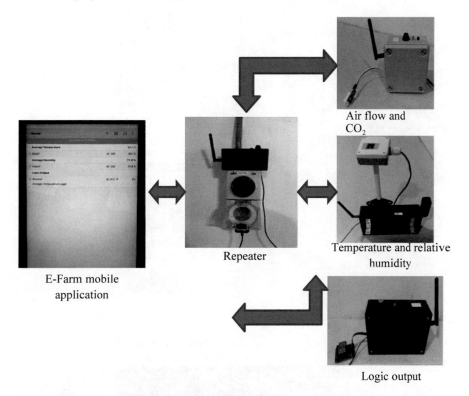

Air flow and CO_2

Repeater

Temperature and relative humidity

E-Farm mobile application

Logic output

Fig. 4.7 Environment monitoring system using mobile application

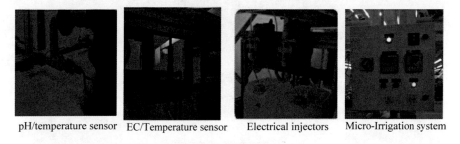

pH/temperature sensor   EC/Temperature sensor   Electrical injectors   Micro-Irrigation system

Fig. 4.8 Micro-irrigation system

pest (whiteflies) on tomato plants. For achieving Agriculture 4.0 system which is constructed from the scratch, such as the robotic harvesting system, MARDI has developed smart machines which integrate high technology robotic system (Fig. 4.11) including a collaborative robot (COBOT), 3D camera and soft grippers to enhance safety and create an efficient working environment. The harvesting process is done without any decision-making by human. Another smart robotics system has been developed in the post-harvesting process, smart vision computer-based robot for

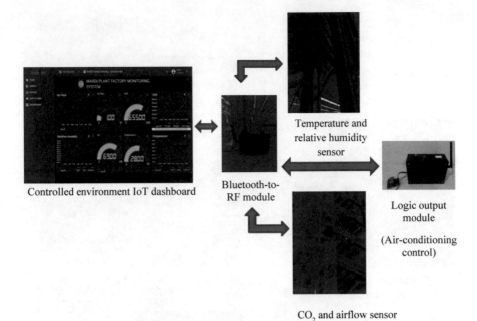

Fig. 4.9 Integration of controlled environment system with IoT application

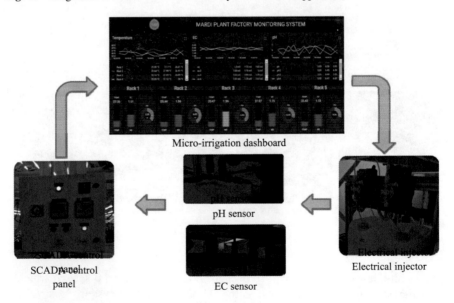

Fig. 4.10 Integration of micro-irrigation system with IoT

Fig. 4.11 Tomato harvesting robot

cleaning edible bird nest. This new innovation machine integrates 6-axis collaborative robot, smart vision camera system and specially designed suction mechanism. This developed machine is capable to clean up to 70% cleanness with 3 times faster than manual method. Agriculture 4.0 also equips the big data technology with blockchain capabilities embedded in the food production system; market actors, authorities and consumers would have access to transparent and decentralized information about food, such as its origins, agricultural practices and pricing. This will sustain the agricultural productivities even in the pandemic situation in the future. During the COVID-19 epidemic, Malaysia's agriculture supply chain has been disrupted as a result of the MCO; however, throughout the MCO, approximately 69% of agri-food producers used an e-commerce platform as an alternative channel [13]. Thus, big data technology with blockchain capabilities embedded in the food production system will be the powerful tools to sustain agricultural activities.

Conclusion

Agriculture will be the most important sector needed to be paid more attention, without food, and all other activities will have to be stunt. Agriculture 4.0 era have to be well established by adopting IR4.0 technologies as soon as possible to sustain the agricultural activities. By achieving these, two of the United Nations SDGs, no poverty and zero hunger, only could come true. However, there will be challenges in this journey such as capacity building and the ability of possessing the advanced technologies, especially for the smallholder farmers in the developing countries. However, these challenges could be overcome with the helps from governments, technologies providers, research organizations and corporations to work together in contributing their strength in sustaining the smallholder farming systems. These could be done by technologies' transfer, providing the package services as the alternative of purchasing the technologies, providing supports to farmers for the new technologies adoptions and providing IT infrastructures to reduce the dependency of the labor force and the period of production and increase the productivity and food quality. These will also encourage more youths in the agricultural sector as they will no longer feel it as 3D sector as "dirty, difficult and dangerous". Consequently, food security will be sustained for the fast-growing world population.

References

1. FAO (Food and Agriculture Organization): Trade Reforms and Food Security: Conceptualizing the Linkages. Commodity Policy and Projections Service Commodities and Trade Division. FAO, Rome (2003). http://www.fao.org/3/y4671e/y4671e00.htm
2. Roser, M.: Future Population Growth. Published online at OurWorldInData.org (2013). Retrieved from: https://ourworldindata.org/future-population-growth
3. Gregorioa, G.B., Ancog, R.C.: Assessing the impact of the covid-19 pandemic on agricultural production in Southeast Asia: toward transformative change in agricultural food systems. Asian J. Agric. Dev. 17(1362–2020–1097), 1–13 (2020)
4. De Clercq, M., Vats, A., Biel, A.: Agriculture 4.0: the future of farming technology. In: Proceedings of the World Government Summit, Dubai, UAE, pp. 11–13 (2018)
5. Klerkx, L., Jakku, E., Labarthe, P.: A review of social science on digital agriculture, smart farming and agriculture 4.0: new contributions and a future research agenda. NJAS-Wageningen J. Life Sci. 90, 100315 (2019)
6. Yahya, N.: Agricultural 4.0: its implementation toward future sustainability. In: Green Urea, pp. 125–145. Springer, Singapore (2018)
7. Santos Valle, S., Kienzle, J.: Agriculture 4.0—agricultural robotics and automated equipment for sustainable crop production. In: Integrated Crop Management, vol. 24. FAO, Rome (2020)
8. Grigore, L.S., Priescu, I., Joita, D., Oncioiu, I.: The integration of collaborative robot systems and their environmental impacts. Processes 8(4), 494 (2020)
9. Zhou, T., Hasheminasab, S.M., Habib, A.: Tightly-coupled camera/LiDAR integration for point cloud generation from GNSS/INS-assisted UAV mapping systems. ISPRS J. Photogrammetry Remote Sens. 180, 336–356 (2021)
10. Abu Bakar, B., Ahmad, M.T., Ghazali, M.S.S., Abd Rani, M.N.F., MhdBookeri, M.A., Abdul Rahman, M.S., Abdullah, M.Z.K., Ismail, R.: Leveling-index Based Variable Rate Seeding Technique for Paddy. Precision Agric. 1–8 (2019). https://doi.org/10.1007/s11119-019-096 92-4
11. Vijayakanthan, G., Kokul, T., Pakeerathai, S., Pinidiyaarachchi, U.A.J.: Classification of vegetable plant pests using deep transfer learning. In: 10th International Conference on Information and Automation for Sustainability (ICIAfS), pp. 167–172. IEEE (2021)
12. Kandiah, P.: Image Processing an Emerging Technique for Early Detection of Pest and Diseases (2021). Retrieved from: http://repo.lib.jfn.ac.lk/ujrr/handle/123456789/3589
13. Amir, H.M., Abidin, A.Z.Z., Rashid, K.F.A., Suhaimee, S.: Agriculture Food Supply Chain Scenario during the COVID-19 Pandemic in Malaysia. FFTC Agricultural Policy Platform (FFTC-AP) (2020). https://ap.fftc.org.tw/article/2679

Chapter 5
Improved Efficiency of Smart Agriculture By Using Cost-Effective IoT Design

Than Htike Aung, EiNitar Tun, and Soe Soe Mon

Introduction

In Myanmar, agriculture accounts for a quarter of the country's economy, with 70% of the population living in rural areas [1]. IoT designs need to be developed to enable the application of smart agriculture methods that can be integrated with potential technologies that drive the development of the agricultural sector. If Myanmar's well-supplied agricultural sector can increase productivity and modernize it, it could have the potential to become a major agricultural and food hub in Asia. Myanmar's agriculture sector accounts for 38% of GDP. There is 20–30% of export revenue. As a result, the employment rate is 70% [2]. The traditional cultivation cost is 150,000 kyats per acre, so the profit is very low [3]. If only farmers could use smart farming, market-smart agriculture and climate-smart agriculture and change crops, it can generate more revenue. Agriculture is the basic food source for the human species. It plays an important role in the country's economic development. Farmers are still using traditional farming methods, which have reduced crop yields. In the agricultural sector, modern science and technology are needed to increase yields.

The Internet of Things (IoT) is widely used to connect devices and collect data. It is used to monitor humidity, temperature, soil, rainfall, fertilizer and storage of land in agricultural areas. The use of IoT can increase productivity at a lower cost. Modern farming systems combine traditional methods with the latest technologies, such as the Internet of Things, and connect to wireless sensory networks [4]. IoT technology

T. H. Aung (✉) · E. Tun · S. S. Mon
Ministry of Science and Technology Myanmar, Naypyidaw, Myanmar
e-mail: thanhtikeaung@ucspyay.edu.mm

E. Tun
e-mail: einitartun@ucspyay.edu.mm

S. S. Mon
e-mail: soesoemon@ucspyay.edu.mm

© Centre for Science and Technology of the Non-aligned and Other Developing Countries 2023
K. Pakeerathan (ed.), *Smart Agriculture for Developing Nations*, Advanced Technologies and Societal Change, https://doi.org/10.1007/978-981-19-8738-0_5

is a wireless sensor network protocol that collects data from various types of sensors and sends it to the main server. Data from all these nodes are collected and sent to the cloud storage. Data collected by sensors are stored using cloud services.

Farmers can access their separate accounts to view the current data of each node through their mobile phones. Factors that can greatly affect productivity can include the attack of insects, pests, wildlife and birds. Crop yields may decline due to unpredictable monsoon season rainfalls, water scarcity and improper water use. Therefore, the use of stored information can predict and increase productivity.

Related Work

Gondchawar and Kawitkar [5] proposed IoT-based smart farming. The purpose of this paper is based on smart agriculture automation and IoT technique. Smart GPS-based remote control robot will perform functions such as sensing moisture weeding and spraying. It includes smart irrigation system, decision-making based on real-time field data and smart warehouse management. All functions are controlled by a smart device. Features include sensors modules, ZigBee modules, a camera, microphone controller and a Redberry controller. All sensors and microcontrollers are successfully connected to the three nodes using wireless communication with the Raspberry Pi. This article provides information about field work, smart irrigation system and storage problems by using remote control robots for smart irrigation and smart warehouse management systems.

Rajalakshmi and Devi Mahalakshmi [6] designed to monitor crop field uses soil moisture sensors and humidity sensors, light sensors and temperature sensors. Sensors that collect data use a JSON format data code to send a web server using wireless transmission to maintain it in the server database [7]. This project focuses on the safety and protection of agricultural products from rodent or insect infestation in fields or paddy stores [8]. This paper describes wireless sensor networks. The network collects data such as temperature and soil moisture. It performs two nodes: collection and analysis. Kassim et al. [4] described the work of precision agriculture (PA). WSN is the best way to solve agriculture issues such as improving agricultural resources, decision-making support and land monitoring [4]. This article describes the greenhouse technology. Representative design in agricultural technology is based on the implementation of the CC2530 chip using ZigBee technology.

Proposed System

Figure 5.1 shows the block diagram of proposed system design. In the proposed system, the controller uses esp8266, and the sensors are connected to it. The main components of the sensor are dht11 and soil moisture sensors. The esp8266 module processed the input sensors' value and then sends to cloud storage real-time firebase

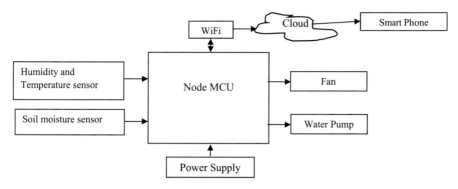

Fig. 5.1 Block diagram for proposed system

database. Farmer can view online agriculture information with their Android smart mobile phone.

Figure 5.2 shows a system flow diagram of the proposed system. When proposed system starts read the data from humidity and temperature and soil moisture sensors, then esp8266 WiFi module processed the input sensors' values. It has two conditional branches; firstly, if input values of humidity and temperature are greater than user define value, then operate fan on, otherwise fan off. Secondly, if soil moisture value is less than user-defined value, hen water pump is on, otherwise water pump is off.

Devices Used

Humidity and Temperature Sensor (DHT11)

Figure 5.3 shows DHT11 sensor module. Humidity and temperature sensor (DHT11) consists of a thermistor, humidity sensing component and an IC. Thermistor calculates the temperature of its surrounding medium from its capability of varying its resistance due to temperature. A moisture holding substrate is placed between two electrodes in humidity sensing component. The variation in humidity produces a variation in resistance between electrodes. The variation in resistance is measured and processed by the IC which gives the humidity value to the NodeMCU. This sensor operates at a voltage range of 3.3–5 V. The range of temperature is 0–50 °C, and range of humidity is 20–90% RH.

Soil Moisture Sensor

Figure 5.4 shows soil moisture sensor module. The soil moisture sensor computes the average of dielectric permittivity along the length of the sensor. The temperature range for the working of this sensor is 10–30 °C, and voltage applied is 5 V.

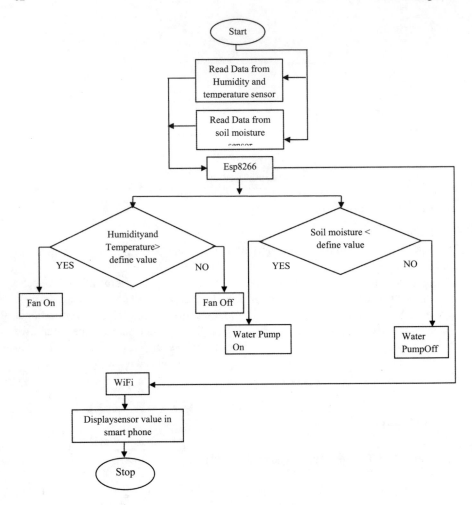

Fig. 5.2 Flow diagram for proposed system

Fig. 5.3 Humidity and temperature sensor module

Fig. 5.4 Soil moisture
sensor module

Fig. 5.5 Esp8266 WiFi
module

NodeMCU (ESP8266)

Figure 5.5 shows Esp8266 WiFi module. NodeMCU is an open-source IoT platform which includes firmware that runs on ESP8266 Wi-Fi module. Programming is done in Arduino IDE using C/C++ language or Lua script. NodeMCU has 16 GPIO pins which can be used to control other peripheral devices like sensors, LEDs, switches, etc. These pins can also be used as PWM pins. It has two UART interfaces and uses XTOS operating system. It can store 4 M bytes of data. The operating voltage of NodeMCU is 5 V. It uses L106 32-bit processor, and the processor's speed is 80–160 MHz.

Display Screen

Figure 5.6 shows IoT OLED screen display module. It is 128 × 64 0.96-inch OLED display screen which offers a stylish interface for proposed IoT projects. The OLED display using the SSD1306 processor communicates with the I2C protocol. On this screen with four pins, vcc is + 5 V, gnd is ground and sda and scl are i2c pins.

Fig. 5.6 IoT display screen module

Power Supply

Figure 5.7 shows power supply module. Breadboard Power Supply Module 3.3 and 5 V Better Version with Micro-USB Interface Power Supply Voltage Regulator 3.3 and 5 V, 1 A for Prototyping Breadboards. This power supply board needs an input voltage of 5–12 V DC (maximum 15 V), like a 9 V battery. It has 3.3 and 5 V outputs, selectable for each of the two power lanes. So, you can have 3.3 V on one lane, and 5 V on the other one. Extremely convenient if you work with mixed logic level devices. This breadboard power supply module comes with a mini USB input, instead of the cumbersome USB-A type of the older version.

The maximum output current per voltage is 1 A (combined over both lanes). The output current depends on the input voltage and can be up to 750 mA at 3.3 V and 1.5 A at 5 V when powered with 5 V 250 mA at 3.3 V, 375 mA at 5 V when powered with 9 V 100 mA at 3.3 V, and 150 mA at 5 V when powered with 15 V. This module fits perfectly on standard solderless breadboards.

Fig. 5.7 Power supply module

Implementation of Proposed System

Figure 5.8 shows agriculture monitoring system device implementation in laboratory. The agriculture monitoring device is very useful for farmers. Their total cost for one device is 27,200 kyats. Therefore, it is a reasonable amount of money that every farmer can use.

Figure 5.9 shows android mobile application. Farmers can use it anywhere in the telecommunications area. They can monitor their farms in real time.

Figure 5.10 shows firebase real-time database for proposed system. Client nodes' input values are stored in firebase real-time database.

Fig. 5.8 Smart agriculture IoT device for proposed system **a** IoT device **b** IoT device with stand

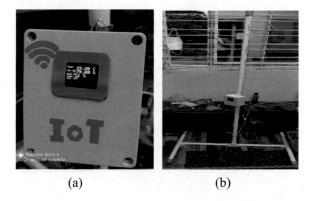

(a)   (b)

Fig. 5.9 Smart agriculture IoT device for proposed system **a** IoT monitor icon **b** IoT monitor dashboard

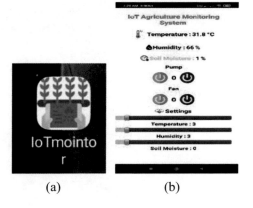

(a)   (b)

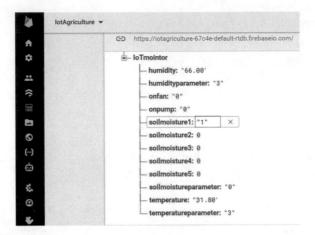

Fig. 5.10 Firebase real-time database

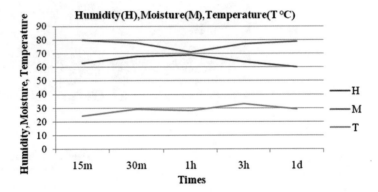

Fig. 5.11 Output results for input devices

Experimental Results

Figure 5.11 shows output results for input devices. The input temperature is normally between 20 and 30 °C. The input humidity is between 60 and 80. The input moisture is nearly 80.

Conclusion

The IoT is vast and can compile a collection of information. The smart agricultural system is a reliable system. It is an effective system that can correct action to the error. Monitoring of farmland saves human resources and improves crop yields. This

IoT monitoring device minimizes power consumption and save costs. Due to the use of solar power, the cost is only the cost of the devices. The proposed system is more efficient and beneficial for farmers. It provides the information of the temperature, humidity and soil moisture in the agricultural sector to farmers monitoring with telephone. Myanmar has over 4 million acres of arable land. If farmer uses this device, they can increase the crop production significantly. If 200,000 acres are used, 479 devices will be enough. For 4 million acres of agricultural land, 9580 units need 26,057,600 kyats. Farmers need to be able to start using this system in the future. The system can be further implemented by adding solar energy and adding the necessary sensors.

References

1. UNFPA: Myanmar Population and Housing Census (2014). https://myanmar.unfpa.org/en/country-profile-0
2. https://www.export.gov/apex/article2?id=Burma-Agriculture
3. FAO: Handbook on Climate Smart Agriculture in Myanmar, 192pp. Nay Pyi Taw (2019)
4. Kassim, M.R.M., Mat, I., Harun, A.N.: Wireless sensor network in precision agriculture application. In: International Conference on Computer, Information and Telecommunication Systems (CITS), Published in IEEE Xplore (2014)
5. Gondchawar, N., Kawitkar, R.S.: IoT based smart agriculture. Int. J. Adv. Res. Comput. Commun. Eng. **V**, 838–842 (2016)
6. Rajalakshmi, P., Devi Mahalakshmi, S.: IOT based crop-field monitoring and irrigation automation. In: 10th International Conference on Intelligent Systems and Control (ISCO), pp. 7–8 (2016)
7. Baranwal, T., Nitika, Pateriya, P.K.: Development of IoT based smart security and monitoring devices for agriculture. In: 6th International Conference—Cloud System and Big Data Engineering. IEEE (2016). 978-1-4673-8203-8
8. Sales, N., Arsenio, A.: Wireless sensor and actuator system for smart irrigation on the cloud. In: 2nd World forum on Internet of Things (WF-IoT), Published in IEEE Xplore Jan 2016 (2015). 978-1-5090-0366-2

Chapter 6
Smart Greenhouse Development: A Case Study in West Java, Indonesia

Sulastri

Introduction

The food demand has increased in the recent decade due to the accelerated population growth and economy. The agricultural sector has an important role in improving the economy and to meet food needs. Besides being known as a maritime country, Indonesia is also known as an agricultural country because most of the Indonesian population has a livelihood as farmers or farming [12]. Statistics Indonesia (BPS) released that the GDP growth of the national economy in the second quarter of 2020 decreased by 4.19% (Q to Q) and year on year (YoY) decreased by 5.32%. The agricultural sector is still the largest contributor to the national GDP. Agricultural GDP grew 16.24% in the second quarter of 2020 (Q to Q) even on a YoY basis, the agricultural sector continued to make a positive contribution, which grew by 2.19% [14]. Nevertheless, there are many challenges in agriculture production. Despite the fact that Indonesia has human resources and a large amount of land suitable for the development of agriculture, food production is being affected by climate change.

Several issues of the agriculture sector regarding climate exchange are identified as follows:

1. Agriculture sector is the main producer of food, supplier of agro-industry, and bioenergy;
2. Sea level rise would decrease agriculture land in the coastal zone;
3. Increase of atmospheric temperature would decrease crop productivity, damage agriculture land resources, and infrastructure;
4. Limited land resources because of degrading land quality and declining production potential; and

Sulastri (✉)
Center for Applied Microbiology, National Research and Innovation Agency (BRIN), LAPTIAB Building, Kawasan Puspiptek, Serpong, Tangerang Selatan, Indonesia
e-mail: sulastri.3@brin.go.id

© Centre for Science and Technology of the Non-aligned and Other Developing Countries 2023
K. Pakeerathan (ed.), *Smart Agriculture for Developing Nations*, Advanced Technologies and Societal Change, https://doi.org/10.1007/978-981-19-8738-0_6

5. Change in rainfall pattern, causing a shift in planting period, season and planting
    pattern, land degradation, and decrease in water availability [1], pest and disease
    problem [13].

Furthermore, as a result from many studies has indicated that climate change
could result in a 9–25% reduction in farm level net revenue in Indonesia in the future
[3]. Conversion of agricultural land to other uses also contributes to the reduction of
arable land. Over the past two decades, around half a million ha of irrigated land have
been converted to other uses; an average of about 25 thousand ha per year converted
to non-agricultural use occurs in Java, Indonesia's most populous island [5].

One of the factors that have most affected the agricultural sector in recent years
is the uncontrolled increase in temperature, which directly affects crops, livestock
production, and health of the population. Agricultural food production is also vulnerable to temperature increase. This is because plants need a certain range of climate
(temperature, precipitation, etc.) for optimal growth and harvest. The decrease in
planting area caused by an increase of temperature in 2050 is predicted to reach
3.3% in Java and 4.1% outside of Java [6]. Given favorable climate and fertile land,
West Java become the most densely populated province and as one of the largest
producer of rice and horticultural products in Indonesia. Currently agriculture food
production in West Java is also facing same problem as mentioned above regarding
reduction of arable land. Traditional cultivation is accordingly altered to an intensive
use of land with modern agricultural technology to meet the increasing food demands.
Crop development in greenhouses or protected cultivation can be a solution to meet
this problem. This controlled environment and resource-intensive agriculture offers
high quantity of quality production thus can provide attractive yields for farmers [10].
In this paper, the current status and challenges regarding the implementation of smart
greenhouse in horticulture production in West Java, Indonesia, will be presented. The
influence of the recent development of smart greenhouses technology will also be
outlined.

Horticulture Productions in West Java

The West Java regency has high production of horticulture especially for seasonal
vegetables (chili, paprika, cayenne pepper, potato, tomato, carrot, lettuce, and string
bean), cut flower (rose, gerbera, and crissantium), and other ornamental plants and
fruits (strawberry, citrus, lemon, banana, and pineapple). Referring to data from the
Statistics Indonesia (BPS), West Java was ranked first as an exporting province in
Indonesia with an export value of January–September 2020 reaching 19.11 billion
US dollars or 16.31% of Indonesia's total export period. The export transaction
value of the agricultural sector throughout 2020 for vegetable commodities reached
2,774,054 kg and the transaction value reached Rp 46 billion with destination countries China, Thailand, and Singapore. Fruit exports reached almost 2,540,961 kg (Rp
44 billion) to Singapore, Malaysia, China, and Hong Kong, medicinal plants reached

10,800 kg (Rp 312 million) to Singapore, and ornamental plants 74,520 stems (Rp 2.49 billion) with 54 destination countries such as Turkey, the USA, Japan to Italy [14].

The cultivation of vegetables is conducted in open field (Fig. 6.1) and in the greenhouse (Fig. 6.2). The potential of greenhouse cultivation for chili, paprika, tomato, and lettuce was nearly double compared to open field cultivation. The energy output–input ratio was higher in greenhouse cultivation. Except the production of lettuce that ratio in open field production was only half from that in greenhouse production [10]. In the vegetable production in West Java, some obstacles were identified like insect, diseases, quality of infrastructure, seed shortage, fertilizer shortage, pesticide, capital, labor, irrigation, hydroponic substrate/media, fluctuation price, technical information, market info, storage to keep from damage, and fresh [11].

Fig. 6.1 Typical of open field cultivation of vegetables in West Java

Fig. 6.2 Typical of protected cultivation (traditional greenhouse) in West Java

Smart Greenhouse Technology Developments in West Java

Protected cultivation, namely greenhouse is a closed growing system that maintains climatic variables according to the plant life cycle to stimulate plant development and ensuring production throughout the season [4]. However, smart greenhouses can be considered as a new technology in the agricultural sector, which can contribute to the rapid conversion to sustainable and precision agriculture and the modernization of the sector by utilizing advanced control, measurement and communication infrastructure techniques, and intelligent management solutions. Each greenhouse has cooling/heating system, ventilation and air conditioning units, pumps, fans, CO_2 generators, artificial lighting, sensors, management units, and information and communication infrastructure [2]. West Java province as the most productive area for vegetable cultivation is in the early stage in adopting the smart greenhouse technology in order to improve the quality and quantity of its production. Protected cultivation has been long adopted by vegetable farmer in West Java; therefore, application of adapted technology for protected cultivation will make a contribution for vegetables production.

PT. Nudira Sumber Daya Indonesia (Nudira Farm) located in Pengalengan District, Bandung Regency, West Java Province, is one of the agricultural industries that applied smart greenhouse technology in West Java. Nudira's smart greenhouse is one of the commercial and education farms which has 2600 m² smart greenhouse mainly operated for cherry tomato production. It is equipped with sensors to monitor light, temperature, humidity, and CO_2 as well as to actively control the actuators, namely side screens, ventilation, cooling pads, irrigation, and fertilizer dosing for controlling important plant growth factors (Fig. 6.3). The potential for smart greenhouse cherry tomato cultivation in Nudira's smart greenhouse is seven times that of open field cultivation where the average cherry tomato production is 3 kg/m² in open land and 22 kg/m² in smart greenhouse cultivation. Another commercial company that was developing smart greenhouses are PT. Agro Jabar and PT. Agrindo Karya Persada which is located in Wanaraja District, Garut Regency, West Java Province. PT Agro Jabar together with PT Agrindo Karya Persada developed smart greenhouse base IoT with a drip irrigation system for melons, peppers, cherry tomatoes, beef tomatoes, kyuri cucumber, and lettuce to increase the growth and productivity of cultivated plants.

Furthermore, to accelerate the implementation of digital technology and its integration with rural agriculture, the West Java local government has established a platform to enable quality agricultural products and security traceability, basic agricultural data, and market information of key agricultural products. One of the programs implemented is a land use map information system. Through this application, land use can adjust the type of commodity planted with land conditions and planting time. Another program related to the development of digital agriculture in West Java is the smart greenhouse pilot established by the Ministry of Agriculture (Kementan). The Ministry of Agriculture has officially launched a smart greenhouse pilot for Internet of Things-based horticulture production. An advanced hydroponic system

Fig. 6.3 Smart greenhouse with cooling pad system at Nudira Farm, West Java

where applied and the entire cultivation process is automatically controlled. With this smart greenhouse base IoT system, the productivity of horticultural crops can be maximized. This is because microclimate factors consisting of temperature, humidity, light, and nutrients are optimally controlled and maintained at an ideal level.

Currently, the Ministry of Agriculture has launched six smart greenhouse units built in the agricultural polytechnic development campus (Polbangtan). The smart greenhouse is also used as a learning place for students who are educated to become young farmers after graduation. The results of cultivation in this smart greenhouse are also marketed to off-taker companies with a contract system. The company that has signed the contract is PT. Paskomnas for lettuce and melon commodities, while PT Agrifarm for pokcoy. This smart greenhouse development pattern will ensure continuity, quality, and of course plant productivity so that the results can be adjusted to the demand of off-takers. This smart greenhouse pilot is the result of a collaboration between the Directorate General of Agricultural Infrastructure and Facilities (PSP) and the Agricultural Human Resources Extension and Development Agency (BPPSDMP), in this case the Agricultural Education Center (Pusdiktan) which is committed to producing independent greenhouses, job seekers and job creator, modern and independent, professional, and able to compete in the industrial era 4.0 [8] (Fig. 6.4).

Moreover, in line with those programs, the West Java Provincial Government has prepared agricultural land to be managed by 5000 Millennial 4.0 farmers. They must follow the selection and have the passion to develop agricultural products. This program aims to make the younger generation no longer looking for work in the city. Creating a generation of Millennial Farmers 4.0 to become young farmers is one step so that food remains the main economic power of West Java. The land prepared for this activity is 12 ha, for capital loans will be provided through the People's Business Credit (KUR) and this system has been approved by the bank (https://deskjabar.pik iran-rakyat.com).

Based on the data information obtained, the smart greenhouses that have been developed in West Java are mostly conducted by industries and are still in its early stage. Smallholder farmers are not yet turned to smart greenhouses for horticultural

Fig. 6.4 Smart greenhouse pilot for lettuce cultivation in Polbangtan, Bogor, West Java. *Source* https://mediai ndonesia.com

production, especially for vegetables. Though the concept of the smart greenhouse is not yet adopted on a large scale in West Java, several crop industries which have been implemented this technology have benefitted immensely from its usage. Such modern greenhouse can sprawl for big crop industries and can also be implemented for small-scale farms. Therefore, it is important to support the initiatives program for implementation of smart greenhouse system for smallholder farm. We can nonetheless see that the development of smart greenhouse as even a small part of the development of digital technologies in agriculture for smallholder farm is lagging and faces numerous challenges.

Opportunities and Challenges

The feasibility study conducted by Kuswardani et al. [11] in two districts in West Java was found that cultivation for vegetable production under greenhouse system has great potential for increasing productivity, energy efficiency, and economic analysis. Nevertheless, it requires knowledge, managerial skill, patience, and hard work of a grower, and government/institutional support. This study result also applied to the development of smart greenhouses as well. Another good opportunity with the development of smart greenhouses is the potential for empowering millennial farmers. Statistical data of West Java shows that about 3.2 million populations are farmers and only 30% or 900 thousand of which are youth or millennials. The agricultural sector in West Java has great potential not only in the rice farming sector but other agricultural production sector potentials that millennials can work on. The young generation or what is currently known as millennial youth is a determinant of future agricultural progress. The most important thing is how to provide education and assistance to these millennial farmers, so they can make a breakthrough for competitive farmers and can create and change the order and mindset of the existence of the farmers themselves.

The main difficulties for most greenhouse growers especially from urban area in adoption technology are lack of technological knowledge, insufficient financial resources, and computer/hardware illiteracy [7], and these are also found in the implementation of smart greenhouse in West Java. Other obstacles identified are as follows:

- Need for infrastructure improvement. To implementation, the Internet of Things (IoT) requires good Internet access, meanwhile throughout West Java even Indonesia not all Internet access goes well.
- Budgeting investment. The implementation of smart greenhouse technology is not cheap, especially for the advance tools.
- Human resources. Most of the farmers are not yet technology literate. Although the smart farming development is focused on millennial farmers, the importance of technology also affects non-millennial farmers, because currently there are still many elementary and junior high school graduates who are still in their productive age, they also still play a role in agriculture development. Furthermore, most of the farmers are over 40 years old and more than 70% of farmers in Indonesia including in West Java only have an elementary education level and below.
- Condition of agricultural land. It is undeniable that the distribution of population and development in Indonesia is not yet fully evenly distributed. This is evidenced by the fact that there are still many "abandoned lands" or lands that have not been cultivated by people in remote areas, while land in strategic areas is being contested at high prices. Given the increasing land prices, the area of agricultural land owned by farmers in Indonesia as well as West Java is on small average [9].

Conclusions

The young generation or what is currently known as millennial youth is a determinant of future agricultural progress. The development of smart greenhouses in West Java is targeted to support the millennial generation who have innovations and creative ideas that are very beneficial for agricultural sustainability. Therefore, in the future, it will be further improved to create professional human resources through education and training on smart greenhouses and other agricultural technologies. This requires strengthening in the research centers and institutions with specialized human resources and equipment to successfully develop an implementation technology. Implementation of smart greenhouse technology has a great potential for vegetable production in West Java, but it's required the accelerating application of Internet of Things, Internet of Intelligence, Big Data, Cloud Computing, and other new generation information technologies which is still lagging and faces various challenges. The political initiatives developed to support the agricultural sector and agricultural digitalization have created the ideal scenario for the implementation of smart greenhouse in West Java, especially for smallholder farm. The development of smart greenhouse and its priority action is allowing the integration of governmental and private sectors for the generation and development of strategies that provides adequate solutions to the problems of its development. It is important to take into account the social and environmental aspects in the implementation of smart greenhouse not only to highlight the technological advances.

References

1. Badan Perencanaan Pembangunan Nasional (BAPPENAS): Indonesia Climate Change Sectoral Roadmap-ICCSR Synthesis Roadmap, 1st edn. Jakarta, Indonesia (2020)
2. Bersani, C., Ouammi, A., Sacile, R., Zero, E.: Model predictive control of smart greenhouses as the path towards near zero energy consumption. Energies 13, 3647 (2020). https://doi.org/10.3390/en13143647
3. Center for Climate Change and Multilateral Policy: Enabling environment for private sector engagement in climate change adaptation projects. Indonesia: NDA GCF Indonesia Secretariate (2020)
4. Cepeda, P., Ponce, P., Molina, A., Lugo, E.: Towards sustainability of protected agriculture: automatic control and structural technologies integration of an intelligent greenhouse. In: Proceedings of the 11th IFAC Workshop on Intelligent Manufacturing Systems, São Paulo, Brazil (2013)
5. Gunning-Trant, C., Sheng, Y., Hamsher, P., Gleeson, T., Moir, B.: What Indonesia wants: analysis of Indonesia's food demand to 2050. In: ABARES Research Report no 15.9, Canberra. CC BY 3.0 (2015)
6. Handoko, I., Sugiarto, Y., dan Syaukat, Y.: Keterkaitan Perubahan Iklim dan Produksi Pangan Strategis: Telaah kebijakan independen dalam bidang perdagangan dan pembangunan. SEAMEO BIOTROP, Bogor, Indonesia(2008)
7. Jensen, M.H.: Controlled environment agriculture in deserts, tropics and temperate regions: a world review. Acta. Hort. 578, 19–25 (2001)
8. Kementan: Smart Green House untuk Percontohan Digitalisasi Pertanian (2020). https://mediaindonesia.com/
9. Kilmanun, J.C., Astuti, D.W.: Prosiding Seminar Nasional Kesiapan Sumber Daya Pertanian dan Inovasi Spesifik Lokasi Memasuki Era Industri 4.0
10. Kuswardhani, N., Soni, P., Shivakoti, G.P.: Comparative energy input–output and financial analyses of greenhouse and open field vegetables production in West Java, Indonesia. Energy 53(2013), 83–92 (2013)
11. Kuswardhani, N., Soni, P., Shivakoti, G.P.: Development protected cultivation in horticulture product: feasibility analysis in West Java Province. Int. J. Adv. Sci. Eng. Inf. Technol. 4(5) (2014). ISSN: 2088-5334
12. Nurbaya, S., Hendroyono, B., Mahfudz (eds.): The State of Indonesia's Environment. Ministry of Environment and Forestry, Republic of Indonesia (2020)
13. Pakeerathan, K.: Molecular tagging of disease resistance in oilseed crops. In: Prakash, M., Gomathi, R., Basu, P.S., Vanaja, M., Kalarani, M.K. (eds.) Physiological Interventions for Developing Climate Resilient Pulses and Oilseed Crops. International Books & Periodical Supply Service (Publisher of Scientific Books), Delhi, India. Program Petani Milenial 4.0 Direncanakan Dimulai Awal Maret 2021. (2022). https://deskjabar.pikiran-rakyat.com/
14. Statistics Indonesia: Economic Growth of Indonesia Second Quarter (2020). https://www.bps.go.id

Chapter 7
Maucrop: An AI-Driven Interactive Mobile Application to Advice on Crop Selection and Cultivation for Small-Scale Crop Farmers in Mauritius

Sandhya Armoogum, Geerish Suddul, Girish Bekaroo, and Aditya Santokee

Introduction

Agriculture and farming are one of the oldest and most important professions in the world. In the past, the agricultural sector has played a pivotal, economic role and served as a driver in the development of Mauritius. Since the early 1970s, however, the contribution of agricultural production to Gross Domestic Product (GDP) has been declining steadily from around 30% to only 3.4% in 2013, largely as a result of the successful diversification of the economy into the manufacturing and services sectors [1]. Out of these 3.4%, some 2.2% are generated by the sugar subsector. The agriculture sector has not only faced a decrease in GDP contribution but also a slight decrease in land utilization and in employment [1]. Nonetheless, the sector still plays a vital, multi-functional role within the economy as it provides direct employment to some 44,200 persons. The agricultural industry also has social and environmental impacts.

Food crop production is dominated by small-scale farming and covers a wide range of crops including potatoes, onion, tomatoes, chillies, crucifers, cucurbits, leafy vegetables, garlic, and ginger which are cultivated on commercial scale. Although there are a few irrigated networks, food crop production continues to be largely

S. Armoogum (✉) · G. Suddul
University of Technology Mauritius, Pointe-Aux-Sables, La Tour Koenig, Port Louis, Mauritius
e-mail: sandhya.armoogum@utm.ac.mu

G. Suddul
e-mail: g.sudddul@utm.ac.mu

G. Bekaroo · A. Santokee
Middlesex University (Mauritius Branch Campus), Flic en Flac, Mauritius
e-mail: g.bekaroo@mdx.ac.mu

A. Santokee
e-mail: a.santokhee@mdx.ac.mu

© Centre for Science and Technology of the Non-aligned and Other Developing Countries 2023
K. Pakeerathan (ed.), *Smart Agriculture for Developing Nations*, Advanced Technologies and Societal Change, https://doi.org/10.1007/978-981-19-8738-0_7

rainfed, resulting in surplus vegetable production during the winter months and a shortage in the summer months. Over the last decade, production of selected crops such as tomato, green pepper, and cucumber has started under soil-less, protected structures. There is also growing interest for mushroom production. Generally, some 8000 small producers cultivating about 8200 ha of land produce, on average, some 110,000 tonnes of food crops annually. Previously, production adequately satisfied the local consumption except in cases of drought, cyclones, and heavy rains. Still, Mauritius is a net food importer, with imports close to 77% of its food requirements. Indeed, imported agricultural products were valued at MUR 36.4 billion compared just over MUR 23.6 billion for exports. Main items imported include wheat, rice, oil, fresh fruits, meat, and milk [2]. Over the last 5 years, this gap has been rising, indicating an increasing dependency on imported food even for fruits and vegetables.

According to the "Strategic Plan (2016–2020) for the Food Crop, Livestock and Forestry Sectors" report [3], the following key areas need urgent attention in order to boost the agricultural sector have been identified:

- Increase investment in modern and innovative agricultural practices;
- Promote sustainable agricultural growth;
- Improve agricultural diversification and productivity to ensure further food security;
- Empower individuals further, particularly the youth and women, to undertake agricultural activities; and
- Strengthen agricultural exports, research and development, and capacity building.

Research Aims

As a Small Island Developing State (SIDS), Mauritius is one among the countries that are most affected by climate change and its impacts [4]. According to the Mauritius Meteorological Services [5], a rise in temperature, an increase in the number of consecutive dry days, decrease in the number of rainy days, and numerous flash floods, among others, have been registered during recent years. All these factors have caused a rebound effect on crop selection, sowing and growth [6]. This also makes the selection of plants for sowing and cultivation a more challenging process as inappropriate selection leads to waste and improper growth while also affecting the yield. Plant selection for cultivation depends on various factors including region, temperature, and humidity, among other variables [7]. It is even more challenging to do effective crop selection because of the micro-climates in Mauritius. There are no specific dedicated regions of land for cultivation, but rather food crops are grown all over the island.

As the world population continues to grow, and land becomes scarcer, people have become more creative and efficient about how to farm, using less land to produce more crops and increasing the productivity and yield of those farmed lands. One recognized solution to manage crop selection and production inputs in an environmentally friendly way involves the adoption of precision agriculture [8]. Although precision agriculture could improve crop selection and sowing, limited work has been undertaken in Mauritius in this direction. Furthermore, no survey has been conducted in Mauritius to assess whether farmers within the island are using precision agriculture tools and techniques for crop cultivation. To the best of our knowledge, no ICT tool has been developed to help small farmers manage their farming activities and effectively select crops for sowing and cultivation.

Effective selection of crops for sowing and cultivation is of utmost importance in order to avoid waste, improve growth and yield. However, this depends on various factors where some include soil type, temperature, and humidity, among other variables [7]. Such factors are also country specific. Although some countries including UK [9] and India [10] have already implemented similar tools, these tools cannot be utilized in Mauritius because of country specificity. Additionally, there has been limited application of Artificial Intelligence (AI) techniques within the context of agriculture in Mauritius. To address this gap, this project aims to assess the effectiveness and acceptance of the implementation of an AI-driven mobile-based agriculture tool to recommend on selection and sowing of crops for cultivation within Mauritius.

Background Study and Related Works

Impacts of Climate Change on Agriculture in Mauritius

Agriculture in Mauritius is highly vulnerable to climate change. The effects of global warming, such as higher temperatures during summer, is observed, as well as sea level rise, which leads to saltwater intrusion into the coastal regions of the island. Mauritius is in the cyclonic belt, and the frequency and intensity of cyclones have increased tremendously in the past few years. Rainfall is more unpredictable leading to severe droughts and flash floods. Agriculture in Mauritius is mainly rainfed for small planters in the humid and sub-humid areas, and thus, it is highly impacted by climate changes.

Global warming and climate change are one of the leading factors in deteriorating the optimum conditions for growing crops [11]. For instance, increasing temperature is changing the crop phenology, i.e. the different stages of growth such as flowering and fruiting of different crops are changing. Mahato [12] and Moore et al. [13] concluded that with climate change, mainly variation in temperature, the duration of growing season will be subject to change and will ultimately lead to a decreasing yield percentage. It is expected that with the imminent high temperature, the timeframe for the growing season will decrease, and this may also affect the amount of crops being harvested [14]. According to Sultan [15], the agriculture sector responds negatively to changes in mean summer temperature and precipitation in Mauritius. Thus, small farmers are facing the challenge of adjusting their planting and harvesting dates as well as planning their crop cycle. Moreover, climate change also impacts Mauritius socially and economically. The accelerating rise in sea level and salt water intrusion may affect vital infrastructure as well as reducing coastal land for agriculture.

The tourism, fisheries, freshwater, and health sector are also concerned [16]. Salt water intrusion decreases the amount of freshwater along with the general trend of decreasing rainfall and more frequent droughts which has a direct impact on crop production and livelihoods of small planters. This also results in scarcity of water for consumption. Similarly, climate change affects fish population as well as beaches which attract tourists. With recurrent floods, higher average temperature and higher humidity, Mauritius being a tropical island may also be at risk to vector-borne and infectious diseases.

Furthermore, due to the topology of Mauritius and the increasing construction and urbanization, the micro-climate at different cultivation area varies. Micro-climate is a result of the interaction between the local topography, landscape characteristics, and the regional climate, whereas macro-climate is the climate of a larger area such as a region or a country. Several factors can affect the micro-climate such as the physical features (trees can provide shade, water can provide a cooling effect and hill tops can be windy), shelter (trees, hedges, walls and buildings can provide shelter, which means the area can be warmer), buildings (buildings give off heat that has been stored during the day and also act as a wind break), and the colour of the earth affects the heat absorption capacity.

It has often been noted that cultivating plants under favourable conditions tend to result into a really good harvest while, under some unfavourable circumstances, both quality and quantity are heavily affected [17]. The same can be denoted in Mauritius, as amount of food imported has been increasing year-on-year. For example, value of imported vegetables and fruits increased from around Rs. 3000 million in year 2014 to Rs. 3700 million in year 2018, which is about Rs. 700 million increases in only 4 years' time [2]. These numbers suggest a decrease in overall food production in the island, which implies that more resources need to be spent to import food. Therefore, the ripple effect of this is that the consumers must pay a premium price for imported food.

Cerri et al. [18] expressed in their article that "In terms of annual crop production, the effects of the high temperatures are negative". This result is based on data that was gathered in a tropical region. Similar results are expected due to the tropical weather of Mauritius. "Crops grown in the tropics, especially wheat, exhibit immediate yield decline with even the slightest warming" [11]. From this observation, it can be denoted that most crops exhibit a remarkable drop in quantity grown in general due to increase in temperature. Therefore, these key factors should be closely monitored to get better predictions of potential crop growth. For example, after reviewing weather data from Mauritius Meteorological Services the trends reveal that the annual mean temperature has increased by 1.1° since 2010 [5]. Additionally, many cases reported in the news showcase that excessive rainfall impacts crop quality. Specifically, in conditions where extreme temperature or precipitations are experienced, crop growth and yield is drastically reduced. Another side effect of such weather conditions is that weed and pests invade fields and regress growth of crops. Furthermore, cucurbits crop, for example, undergo a chemical change in unusual warmer and humid weather conditions, which make them inedible.

With various climate changes happening, there is a need for a shift towards effective crop production strategies. A previous review showed that due to climate change, developing countries will be facing losses of 12–14% annually [19]. In the same work, it was noticed that from 1970 to 1990, climatic phenomenon such as typhoons, floods, and droughts have brought losses 82.4% in Philippines total rice production. This highly affects the country's economy, causing welfare losses in both rural and urban households.

An inevitable approach to such issues would be to adopt the use of ICT in agriculture. ICT has been integrated in almost every part of a normal life and is a core fragment of various departments in both the public and private sector. With available technologies, it is not difficult to develop the framework to support farmers in optimizing their harvest, but the real challenge is in making people participate and adopt it. Awuor et al. [20] have proposed that collecting data from various sources to develop a model will allow solving these problems.

The Need for Effective Crop Selection and Production Strategies

With the industrial revolution, most fields and lands are being developed into cities or workplaces; thereby, there is less land accessible for small-scale planters. The same is confirmed in the Voluntary National Review Report of Mauritius (2019), where it is stated that the contribution of agriculture to GDP has decreased to 3 per cent in 2018. The same report is quoted saying, "We import 77% of our food requirements, exposing us to international pressures" [21].

It is also a known fact that after long terms of constant rain and strong gusts, the cultivation in Mauritius suffers a lot to such a point that basic crops such as potatoes and tomatoes are destroyed, and same has to be imported from different countries [22]. This results in a spike of up to 150% in the price due to shortage in supply for food. The crops on sale at the time also suffer in quality. Farmers are impacted financially because they are not able to obtain a return on their investment. There are also additional costs to clear the land and start growing new crops again.

Kurukulasuriya and Mendelsohn [23] conducted a study on adapting crop selection to the climate change in Africa. The study involved 11 countries from the African continent. The study concluded that crop selection process is dependent on climatic behaviour of the region. To maximize the profits, one will want a high yield rate, and with an optimal maintenance cost. Hence, the farmer will choose a crop which is favourable to the location's climatic condition, to increase the chances of high yield. In this study, among the countries participated, in the hottest region, farmers tend to plant cowpea and millet, maize–beans, and sorghum in the coolest region, millet and sorghum in the dry regions, and maize–beans, cowpea–sorghum, and maize-groundnut in the wet regions. Maize was found to be cultivated in various regions. In the course of the study, although monthly meteorological data was available, it is suggested to aggregate the data into seasons before proceeding with correlating the crops' harvest with the climate.

A similar study was conducted with respect to the South American farms [24]. The research was based on the hypothesis that the aim for farming is to end up with a significant profit; hence, an effective crop selection is a determining factor for profit optimisation. A more rigorous study was conducted to examine whether the choice of crops is affected by climate in Africa involving 7000 farmers across 11 countries in Africa. It was concluded that crop choice is very climate sensitive [25]. Thus, it becomes important to investigate the influencing factors for more effective crop selection.

Factors Influencing Crop Selection and Cultivation

A manual was set by Food and Agriculture Organization of the United Nations (FAO) for farmers, to enable them into making a better choice for their crop to be planted. Similarly, in Mauritius, the FAREI has the "Guide Agricole" which guides planter in choosing the appropriate crop to learn about the duration of the growing stages of the crop and its requirement throughout, such as irrigation and fertilizers more suitable for the different crops [26]. Profit can also be maximized if the water supply and fertilizers are properly regulated, and best possibly without any artificial intervention.

Geographic regions can be categorized as climatic zones, with respect to their mean temperature and precipitation, namely desert/arid, semi-arid, sub-humid, and humid. The water needs for the crop throughout its entire life cycle also need to be known. This allows better planning and timing for sowing the crop. Knowing the crop's soil nutrient requirements facilitates the process. Matching the soil's composition with the crop's requirements naturally reduces the cost for fertilizers and, hence, increases the yield.

Other factors affecting the cultivation process are the weather, season, region of plot of land, crop variety, and market need. These factors can be encapsulated into two major entities, namely internal factors and external factors. Internal factors mainly refer to the heredity or genetic configuration of that particular crop to be cultivated. Attributes such as the yielding capability, chemical structure and composition, growth/maturity time, resistance against pests, flood, drought and salinity, and quality of seeds/grains, and overall quality of the crop defines its genetic structure, and they are less influenced by external factors such as the environment.

In their book, "Concepts of Agronomy" [27], the most prominent factors affecting crop growth in external factors are identified:

- Climatic: Most of the yield of a crop is dependent on the climate that it is in. The variables affecting the same are temperature, amount of sunlight, wind, precipitation, atmospheric pressure, gases, and humidity.

- Edaphic: Edaphic relates to content incoming from soil; therefore, here the variables affecting the crop are the soil living ecosystem (worms, snails, pests), humidity levels of the soil, the amount of minerals and nutrients in the soil, and finally alkalinity or acidity (pH) of the soil.
- Biotic: Biotic factor refers to the outcome either positive or negative resulting from neighbouring living organisms such as plants and animals. Positive outcomes can be snails and insects which help in decomposing organic matter around crops and the negative outcomes can be parasitic weed inflicting high yield losses in crops.
- Physiographic: Physiographic mainly refers to the physical setup of the location, such as the type of the surface of land (steep, level), the light exposure, or elevation with reference to sea level. These conditions would define what type of crop can be grown in that location and land. For example, tea grows best in loamy soil over hill slopes.
- Socio-economic: Socio-economic factors relate to the society preference towards farming in a big organization or working in small scale. Furthermore, the market need is also a big factor which should be satisfied while keeping in mind the crop yield.

Choosing a crop for sowing is not a random decision, but rather premeditated. There can be several reasons which lead to the ultimate decision. The reasons may be personal, financial, climatic or historical in nature. A common example is the cultivation of rice. A person situated in a region deprived of constant rain will definitely not opt for rice since it requires marshy lands. Sometimes, a farmer may also choose a crop which he will be able to harvest at a much faster rate, rather than growing food crops whose life cycle is relatively long whereby it will take him longer to recover his capital.

Methods Used for Crop Selection by Small Planters in Mauritius

Many small-scale planters in Mauritius do not use the smart phones or the Internet to its full potential and rather opt for old and outdated methods for cultivating crops. This was confirmed through a research article published for Mauritius which stated that the majority of small-scale planters have 15 plus years of experience; thus, they are more leaned towards the traditional way of cultivation [28]. The methods used by most small-scale planters include:

- Shared knowledge gained from fellow planters: This method is the practice of working and sharing tips and techniques to grow crops in an optimum way in the planter's community. Many planters also still follow techniques that their ancestors used. The problem with this method is that they are often outdated and do not provide enough crop yield as when using modern agriculture techniques.

- References from old books and online materials: Many small-scale planters refer to old agricultural books and online materials. This can be a good thing; however, most agricultural books are a generic way of learning simple plantation techniques for home use, and books that provide deep insight into agriculture in bulk and its procedures are mostly based on results obtained from experiments performed in optimal conditions. Therefore, to maximize crop yield, this is not the recommended method, especially for small-scale planters where resources are limited.
- Courses offered by agricultural institutions via the government: The government offers many courses on agriculture which are organized by the Food and Agricultural Research and Extension Institute (FAREI) to promote small-scale plantation and enhance import substitution.

Mobile Applications as a Tool for Smart Agriculture

Among the different ICT tools that can be easily adopted is the mobile phone. There is a global trend towards the increasing number of users connected to the network via mobile devices. According to the ITU [29], there are 107 mobile cellular telephone subscriptions per 100 inhabitants in 2018, which implies that today there are more active mobile devices than people in the world. Due to the affordable cost and availability of smart phones, the latter have become a ubiquitous part of everyday life, even in developing countries. Although, mobile phones are primarily used for communication, social networking, and web browsing; smart phones are also equipped with a plethora of sensors that can allow smart phones to capture data in the surrounding environment of the user. Globally, the information and communication technology industry has been flourishing with mobile applications to promote access to information and enhance service delivery. There has been an increase in number of mobile applications for agriculture.

Farmers in developing countries are actively adopting smart phones, and thus, mobile applications that can meet their needs can easily be implemented and made accessible to them. In recent years, lots of applications have appeared for individual farmers. The goal of such mobile applications is to provide farmers with valuable information that can help them improve their planting, cultivating, and harvesting. This information may include agricultural best practices, weather forecasts, and data on disease epidemics. Such applications can also train new farmers and allow experienced farmers to apply new methods. Such mobile applications can also provide information about agricultural products such as seeds, fertilizers, pesticides, etc. Farmers also need some information about the market such that they are able to compare the prices of different crops in the market place and get insights of the actual supply and demand state of a particular crop.

Fafchamps and Minten [30] discussed the advantages of SMS-based agricultural information, such as weather and market information to Indian farmers. Research showed that the level of awareness of e-Agriculture usage is low and that small farmers have to be educated about the different technologies which could greatly enhance their productivity [31]. Similarly, Zakar and Zakar [32] emphasizes on the importance of the use of ICT to give information to planters about market prices, weather, how to select the appropriate fertilizers, quality of seeds, pesticide, intermixture of cropping, water administration, land preparation, harvesting, and the overall adoption of present information technology to improve agricultural practices. The role of using mobile phone among Iraqi Farmers for Agriculture Development is discussed in Glood et al. [33].

Launched in June 2019, the mobile application, "Mokaro", which a first-of-its kind, aims to guide planters in their agricultural ventures. It is developed by the Ministry of Agro-Industry and Food Security, the Ministry of Technology, and Communication and Innovation in collaboration with the FAREI. The "Mokaro" application will serve as a tool for farmers and planters to better plan their agricultural activities and manage resources as well as minimize losses. "Mokaro" application will advise them on future plantation, irrigation, and other field activities following crop analysis and assessment (supply and demand). Farmers will also receive information related to climatic conditions of Mauritius, current agricultural news, and alerts that may be useful for agricultural production and management. Additionally, the application will comprise a feature on planned calendar of activities and events which will inform planters on activities of stakeholders where they may be invited to participate. The tool will enable planters to locate and communicate directly to suppliers and service providers of fertilizers and pesticides as and when required. Input suppliers, service providers, and operators may register themselves for easy access by farmers. Figure 7.1 depicts some of the user interfaces of the Mokaro app.

Fig. 7.1 Mokaro mobile application interfaces

Design and Implementation of "MauCrop" Mobile Application

The "Guide Agricole" from the FAREI has been instrumental for information regarding crops planted in Mauritius. The guide gives detailed information about the different crops, their sowing, planting, and growing phases as well as their requirements such as water and fertilizers. Information from the "Guide Agricole" being too detailed, it was impossible to transfer all the material in the database in this project. Furthermore, as this project is implemented as a pilot to depict the mobile application (MauCrop) functionalities, only a few crops were considered namely tomato, carrot, garlic, aubergine, and cabbage. The chapter focuses on the different stages of the MauCrop mobile application which will be transformed from a design state to executable state.

"MauCrop" Mobile Application

The mobile application is designed to be an aid for small farmers, referred to as the user of the mobile app. It might be typically useful for young farmers who may not be very experienced, but also a support for experienced farmers. For an inexperienced planter, the crop selection feature will be paramount in helping the planter to select which crop is best suited for his/her plot based on the location, soil type, rainfall region, etc. The mobile application is designed to implement the following features:

- The user can register with the mobile app, providing some basic information such as his name, address, age, and experience level as a farmer.
- The user can register his/her plots details (location, size of plot, soil type, etc.) on the app. This allows the app to be customized for the user.
- The user can search and browse information about different crops such as period for growing the plant, the region most suitable for a crop, the amount of water required by the crop, different types of fertilizers required, and at which period. This information is readily and publicly available from the "Guide Agricole" and online from the APMIS website (http://farei.mu/apmis/) in the Publications section [34]. Such information is to be recorded in a database by the system administrator.
- When the user wishes to start a particular cultivation in a plot of land, the application can suggest a list of crop based on the plot location, soil type, and period of time in the year. This crop selection feature of the mobile app is to be further enhanced based on back-end server data analytics using AI, more specifically machine learning.
- For each crop from the proposed list, the mobile application may also give indication about whether this crop is estimated to be in low, medium, or high abundance at the estimated time of harvest. This evaluation will be possible based on the server side back-end analytics which will be done based on the records of what other small planters are planting in the same period of time. This aspect

of collecting, aggregating, and processing data is termed as **crowdsourcing**. The aim of providing information about the expected abundance of a crop is to ensure an adequate supply and demand balance of the produce on the local market. By recommending that the user sows a crop which will not be in high abundance on the market, this ensures that the market price of the produce will be profitable to the user.

- Weather condition is of prime importance to the user. The mobile app is designed to use a weather API to provide weather information for each plot of land, i.e. weather information is provided in a customized manner to the user such that the user can then take decisions whether to irrigate in some circumstances. Mostly though in the context of Mauritius, small farmers depend on rainfall for the water requirements of their plantations except in cases of severe drought where they may attempt some simple irrigation practices. The weather information is also recorded to be used for the data analytics using machine learning.

Crowdsourcing

In this research work, the data collected from the mobile app includes the duration of crop (from sowing/planting till harvesting), the crop yield, and the type of soil on which the crop was grown as well as other products such as fertilizers that were used. Such crowd sourced data serves two purposes in the context of this project. Firstly, as other users' records which crop they are growing on their plots of different sizes in specific period of time, the back-end server computations can estimate the amount of produce, when the crops would be harvested. Such data allows estimating the amount of the produce in the local market. The mobile application can share such information with the users. However, the choice of which crop to grow depends solely on the user. The mobile application merely makes recommendations to the user based on the analysis of the data gathered. Secondly, the mobile application can also use data collected over a long period of time to make further AI-based recommendations to the user. Such recommendations would be made based on predictions using machine learning algorithms for crop selection. By taking into account the micro-climate of cultivation plots, previous crop yields, weather information, experience level of planters, etc., a prediction model can be obtained. Such a model can then give more accurate crop selection recommendations. Supervised machine learning algorithms were used, namely Decision Tree and Random Forest machine learning as they were found to have better accuracy.

Functional and Non-functional Requirements for the Mobile App

Table 7.1 presents the Functional Requirements of the MauCrop mobile app. Table 7.2 presents the Non-Functional Requirements of the mobile application. The small planter is the user of the mobile application.

Table 7.1 Functional requirements of MauCrop app

	Description	Remarks
FR1	User registration and login to the mobile app	Planter's basic information captured and stored
FR2	User can update profile	
FR3	User updates/resets password	
FR4	User searches/views crop information and fertilizers information	The application shall provide the user all necessary information on food crops, which include crop life cycle, fertilizer's composition, and sowing steps. Such data is stored in the database and is based on information available from the "Guide Agricole"
FR5	User registers his/her plots of land on map	The application must have a map to allow users to pinpoint their plot location and add it to the application database. The plot's coordinates will later be used to monitor climate and perform crop selection techniques
FR6	User views registered plots locations on map. User can edit and delete plots of land associated with his account	
FR7	User interacts with the application to do crop selection and gets a list of recommended crops Planter can further select a recommended crop to get information such as expected yield and estimated period of harvest, estimated supply of crop across the island, and expected profit based on statistical price of the crop produce to be able to take a decision	User provides information such as when the new crop cultivation is to be started, on which of the planter's plot and on how much area of the selected plot. This involves dynamic back-end computations to provide a list of recommended crops to the user based on the suitability of the plot's region for specific crop as per the Guide Agricole and climatic history
FR8	User begins plantation of a new crop (active crop) which involves dynamic back-end computations User edits active crop(s) information, e.g. area of land, fertilizers added, seed/plant costs, labour costs as well as amount of harvested produce when available User can change the status of an active crop to inactive after final harvesting of produce	Estimated time of harvest and estimated amount of produce computed for each active crop Aggregation of estimated amount of produce from different planters for the same active crop computed Cost information is maintained in the database as well as used fertilizer information Back-end updates of expenditures and total crop yield (if produce is several times during the life cycle of the crop)

(continued)

Table 7.1 (continued)

	Description	Remarks
FR9	User can view history of crops grown and plot of land utilization	
FR10	Application sends weather notifications to planter	The back-end server requests for weather information from the weather API for the different active crops and give notifications to the planter. Weather information is also saved in the back-end database for constituting the machine learning data set
FR11	User schedules plantation and fertilization events on calendar User edits and deletes events	

Table 7.2 Non-functional requirements of MauCrop app

	Description
NFR1	System usability—user friendliness that is the system should allow the users to access the functionalities easily
NFR2	Real-time notifications—system will display notification
NFR3	Quality assurance—mobile application handles errors and is robust

MauCrop App Data Model

The data model of the mobile application is depicted in Fig. 7.2.

MauCrop App Use Case

Figure 7.3 shows the interaction between the user (small planter) and the MauCrop mobile app.

System Context Diagram

The context diagram defines and clarifies the boundaries of the software system. It identifies the flow of information between the system and external entities. The entire software system is shown as a single process as shown in Fig. 7.4.

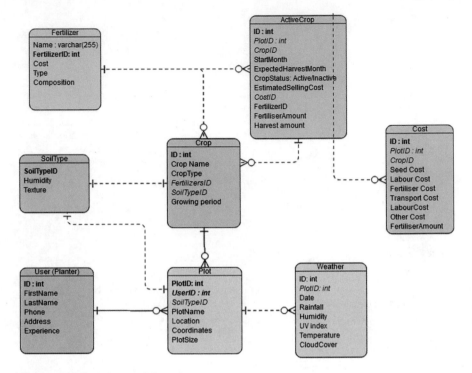

Fig. 7.2 MauCrop data model

MauCrop App Platform

As demonstrated in the context diagram, the system consists of three main components:

(i)   Mobile application: React native.
(ii)   Web services:

    (a) Symphony PHP framework with API platform bundle.
    (b) Flask python framework.

(iii) Persistent storage: MySQL database.

The mobile application communicates with the web services over a network using the standard HTTP protocol. Two web services were developed. The first one is responsible for serving all the transactions that a user initiates on the mobile app, including adding plots, searching information on crops, and starting cultivation on a particular plot. The second one is used to get the humidity region based on the Agricole guide's pluviometry map.

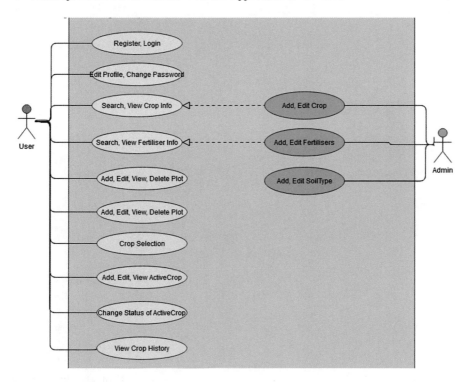

Fig. 7.3   MauCrop use case diagram

Conclusions

Recently, the Smart Agriculture project has been launched in Mauritius, which aims at accompanying the Mauritian agro-ecological transition. The main goal is to promote a well-reasoned mode of production which will allow the general agriculture in Mauritius, to move towards more sustainable and resilient systems in the face of climate change. The MauCrop research project is more focused on the adoption of machine learning techniques to perform better crop selection by small-scale planters of food crops in Mauritius so as to increase crop yield and thus strengthen the resilience of the food production in Mauritius in the face of climate change, declining soil fertility with time, and the reduction of agriculture land due to increasing urbanization. This mobile application would allow small planters to get information about crops and how to go about with the growing/transplanting/planting of vegetables based on the information available from the "Guide Agricole" in Mauritius. This would be most beneficial to young Mauritians or beginners who want to start cultivation of land. The mobile application also allows the small planter to monitor their crops, gets timely weather information for each plot of land, keeps a record of expenses (e.g. seeds, fertilizers, pesticides, labour cost, and travel cost) involved in the cultivation of each crop/plot, and records harvest amount of each crop produce in a timely manner. The

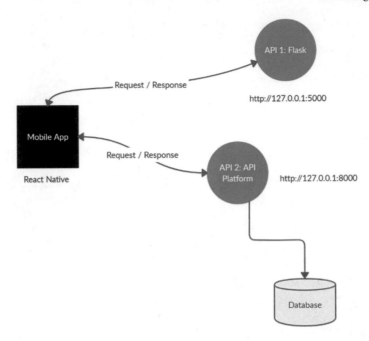

Fig. 7.4 Context diagram

MauCrop mobile application is thus customized for the small planter to help the small planter manage their cultivation of land. The user, when selecting to cultivate a particular crop on a specific plot, can get information such as the expected yield based from both statistical data and the machine learning model prediction and an estimate of the financial gain based on historical sale price of the products at the harvest time right from the mobile app.

Acknowledgements This research work is funded by the Higher Education Commission (HEC) Award Number INT-2018-03 under the Interdisciplinary/Inter-institutional Team-Based Research Scheme.

References

1. Seechurn, et al.: 2013. A Tale of Five Sectors in Mauritius: Agriculture, Textile/EPZ, Tourism, Financial Services and ICT/BPO: An Employment Perspective. Mauritius
2. Statistics Mauritius: Economic and Social Indicators Agricultural and Fish Production. Government of Mauritius, Mauritius (2019)
3. Ministry: Strategic Plan (2016–2020) for the Food Crop. Government of Mauritius (2016)
4. Gray, M., Lalljee, B.: Climate change adaptation in Mauritius: considering the role of institutions. Western Indian Ocean J. Mar. Sci. **11**(1), 99–111 (2012)
5. MMS: Climate Change (2017). [Online] Available at: http://metservice.intnet.mu/climate-ser vices/climate-change.php

6. Belay, A., Recha, J.W., Woldeamanuel, T., Morton, J.F.: Smallholder farmers' adaptation to climate change and determinants of their adaptation decisions in the Central Rift Valley of Ethiopia. Agric. Food Secur. **6**(1), 24 (2017)
7. Barbour, M.G., Burk, J.H., Pitts, W.D.: Terrestrial plant ecology, 3rd edn. Benjamin/Cummings Pub. Co., Menlo Park, CA (1980)
8. Bongiovanni, R., Lowenberg-DeBoer, J.: Precision agriculture and sustainability. Precision Agric. **5**(4), 359–387 (2004)
9. Moonplanter: The Gardeners Calendar (2019). [Online] Available at: https://www.the-garden ers-calendar.co.uk/2019/
10. WSJ: New App Promises to Tell Indian Farmers When to Sow Crops (2016). [Online] Available at: https://blogs.wsj.com/indiarealtime/2016/06/17/new-app-promises-to-tellindian
11. Penman, J., et al.: Good Practice Guidance for Land Use, Land-Use Change and Forestry. Institute for Global Environmental Strategies (IGES) for the IPCC, Japan (2003)
12. Mahato, A.: Climate change and its impact on agriculture. Int. J. Sci. Res. Publ. **4**(4) (2014)
13. Moore, F., et al.: New science of climate change impacts on agriculture implies higher social cost of carbon. Nat. Commun. **8**(1607) (2017)
14. Nelson, G.C.: Climate Change: Impact on Agriculture and Costs of Adaptation. International Food Policy Research Institute (2009).https://doi.org/10.2499/0896295354
15. Sultan, R.: Economic impacts of climate change on agriculture: insights from the small island economy of Mauritius. In Moncada, S., Briguglio, L., Bambrick, H., Kelman, I., Iorns, C., Nurse, L. (eds.) Small Island Developing States, the World of Small States, vol. 9. Springer, Cham (2021). https://doi.org/10.1007/978-3-030-82774-8_7
16. Boodhoo, S.: The Changing Climate of Mauritius, Vacoas-Phoenix, Mauritius, Mauritius Meteorological Services (2008)
17. Johnson, L.K., Bloom, J.D., Dunning, R.D., Gunter, C.C., Boyette, M.D., Creamer, N.G.: Farmer harvest decisions and vegetable loss in primary production. Agric. Syst. **176**(102672), 1–11 (2019)
18. Cerri, C.E.P., et al.: Tropical agriculture and global warming: impacts and mitigation options. Sci. Agric. **64**(1) (2007)
19. Fazal, S.A., Wahab, S.A.: Economic impact of climate change on agricultural sector: a review. J. Transformative Entrepreneurship **1**(1), 39–49 (2013)
20. Awuor, F., Kimeli, K., Rabah, K., Rambim, D.A.: ICT solution architecture for agriculture. In: Kenya, IST-Africa Conference and Exhibition (2013)
21. Ministry of Foreign Affairs: Mauritius to Present its First Voluntary National Review on implementation of SDGs (2019). [Online] Available at: http://www.govmu.org/English/News/Pages/Mauritius-to-present-its-first-Voluntary-National-Review-on-implementation-of-SDGs.aspx
22. United Nations: Policy Makers Digest. United Nations Mauritius, Mauritius (2020)
23. Kurukulasuriya, P., Mendelsohn, R.O.: Crop Selection: Adapting to Climate Change in Africa (2007). [Online] Available at: https://papers.ssrn.com/sol3/papers.cfm?abstract_id=1005546
24. Seo, S.N., Mendelsohn, R.: An analysis of crop choice: adapting to climate change in South American farms. Ecol. Econ. **67**(1), 109–116 (2008)
25. Pradeep Kurukulasuriya, R.M.: Crop Selection : Adapting to Climage Change in Africa. World Bank Group (2013). https://doi.org/10.1596/1813-9450-4307
26. FAREI: Guide Agricole. FAREI, Mauritius (2010)
27. Singh, N.K.R., Singh, A.V.: Concepts of Agronomy, 1st edn. Book Rivers, India (2019)
28. Gadekar, A., Gadekar, R., Panchu, D.: Technology adoption by small planters in Mauritius. Agric. Res. Technol. **2**(2), 555–583 (2016)
29. ITU: ITU Statistics (2018). [Online] Available at: https://www.itu.int/en/ITU-D/Statistics/Pages/stat/default.aspx
30. Fafchamps, M., Minten, B.: Impact of SMS-based agricultural information on Indian farmers. World Bank Econ. Rev. **26**(3), 383–414 (2012)

31. Thankachan, S., Kirubakaran, D.S.: E-Agriculture Information Management System. (2014)
32. Zakar, M., Zakar, R.: Diffusion of information technology for agricultural development in rural Punjab: challenges and opportunities. Pak. Vis. **10**(2), 71–111 (2009)
33. Glood, S.H., Muhil, S., Ibrahim, B.S.: Role of using mobile phone among Iraqi farmers for agriculture development. J. Eng. Appl. Sci. **14**, 5495–5500 (2019)
34. FAREI: Agricultural Production and Market Information System (2020). [Online] Available at: http://www.farei.mu/apmis/

Chapter 8
Affordable ICT Solutions for Water Conservation Using Sensor-Based Irrigation Systems for Use in Arid Agriculture in Thar Desert Region of India

Ravi Bhandari and Anand Krishnan Plappally

Introduction

Agriculture is a service which helps to grow and provide food for human and animal consumption. Water intensive agriculture exerts pressure on water resources. Globally around 72% of fresh water is used for agricultural purposes [1]. Today, the world population is hovering around 8 billion and is expected to reach 10 billion by the year 2050, with agricultural production also expanding by 70% to keep up with the demand [2]. Today, 20% of the total cultivated land is under irrigation, which contributes to more than 40% of the total food produced worldwide, thus indicating that irrigated agriculture is at least twice as productive as other forms of agriculture [2]. Therefore, improving irrigation efficiency is a must for sustainable agriculture and in dealing with a futuristic water crisis.

India is an agrarian economy with agriculture accounting for around 15.4% share in national GDP [3]. The importance of agriculture in the Indian context is reflected by the fact that around 50% of the total population depends on agriculture and allied activities for their survival. Even after 74 years of Independence, the majority of agriculture in India is rainfed. Irrigation only accounts for around 48% of arable land [4] which is less than that of many agrarian economies. With the government planning to increase irrigation coverage rapidly, conventional farming techniques must shift

R. Bhandari (✉)
Department of Computer Science, Indian Institute of Technology (IIT), Jodhpur, India
e-mail: rbhandari@iitj.ac.in

A. K. Plappally
Department of Mechanical Engineering and Center for Emerging Technologies for Sustainable Development (CETSD), Indian Institute of Technology (IIT), Jodhpur, India
e-mail: anandk@iitj.ac.in

© Centre for Science and Technology of the Non-aligned and Other Developing Countries 2023
K. Pakeerathan (ed.), *Smart Agriculture for Developing Nations*, Advanced Technologies and Societal Change, https://doi.org/10.1007/978-981-19-8738-0_8

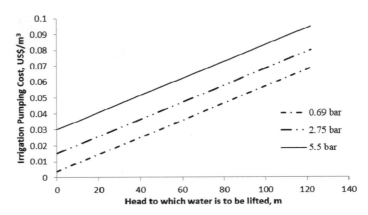

Fig. 8.1 Linear relationships of cost as a function of head at which irrigation is to be performed

to conservation agriculture where resource conservation is key. This demands technological interventions in making irrigation more effectiveness oriented, consuming local produce to reduce virtual water transport, less evapo-transpirasive and less water hungry crops [5].

Gravity-based irrigation is more cost effective than using pumping [6]. The irrigation cost is dependent on the quantity of water, water source, arable land, soil type, landscape, gradient, plants or tree type, regional rainfall, climate, technology used, time, skill sets of farming, pump type and its age, intensity of water application, fuel or electricity usage, and its related cost and water use efficiency [6]. Figure 8.1 illustrates that pumping cost shoots up proportionally with lift required, thus pressurized pumping increases cost.

Information and Communication Technologies (ICT) can prove instrumental in taking a step in this direction, by pursuing precision farming to increase the efficiency of irrigation. For this, the technologies are to be suitably identified according to the requirement for the type of irrigation as illustrated in Fig. 8.2.

Figure 8.2 classified different technologies on the basis of electrical energy intensity for a unit volume of water that passes through it or used by it at a certain operating pressure [6]. Expenses will change as irrigation methods or change with crop requirements. Such intelligent savings of water at the farm level can transform into substantial amounts of water savings on the national and global scale.

Figure 8.3 illustrates that reduction of pumping hours can consequently bring down irrigation costs by using appropriate technology as required by the crop or landscape. These strategies may help reduce the stress on groundwater and freshwater resources near to that landscape [7].

Thus from Fig. 8.1, it is established by various studies that micro-irrigation performs better than surface and sprinkler irrigation for most of the crops [5, 9]. It was important to shed light on the cost and energy expense of micro-irrigation to understand why it was one of the irrigation techniques used and discussed here in this article [6]. Moreover, several studies have demonstrated that micro-irrigation is better suited for arid regions for watering high-value crops, such as fruits, grapes, vegetables, and landscape plants [10–13].

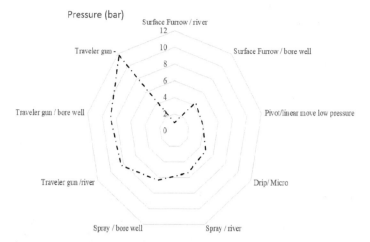

Fig. 8.2 Pressurized-based distinction of irrigation technologies in relation with its water source

Fig. 8.3 Cost of various irrigation technologies while using water from a river source [8]

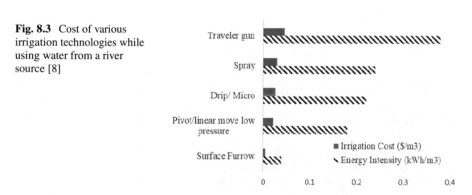

Micro-irrigation or drip irrigation is a method of delivering slow and tiny droplets of water to the soil using a low-pressure distribution system and special flow control outlets. The primary components of a micro-irrigation system are shown in Fig. 8.4. It consists of a control head, main, submains, manifold, and lateral lines to which the emitters are attached.

The control head generally consists of a pump, filters, and flushing valves. It may also have fertilizer injectors, backflow preventers, pressure gauges, pressure regulators, water meter, air relief valves, and programmable control devices. The line to which laterals are connected is called *manifold*. The manifold, submains, and main may be on the surface or buried underground. The lateral lines have diameters generally ranging from 10 to 32 mm and have emitters placed at regular spacing depending on the crop to be grown.

With this system of irrigation, only the root zone of the plant is supplied water which leads to a decrease in evaporation and deep-percolation losses. Labor requirements in contrast to sprinkler-based systems are less, and it can be readily automated.

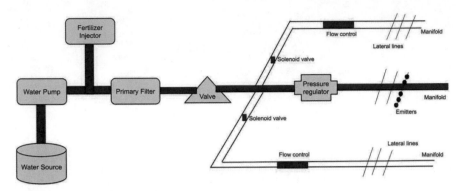

Fig. 8.4 Components of a micro-irrigation system, adapted from Schwab et al. [9]

Bacteria, fungi, and other pests that depend on a wet environment are reduced as the above-ground part is dry, and only the root zone area is wet. Also, liquid fertilizers and pesticides can be applied after mixing thoroughly with water, thus allowing it to reach the roots of the plant, which is not possible with manual sprinkler, surface, or spraying techniques. This makes micro-irrigation a natural choice for our setup.

This paper attempts to establish by field experiments that efficiency of micro-irrigation can be increased with the adoption of ICT techniques, which we believe is a novel attempt in this area and also a major contribution of this paper. Another important aspect is the affordability of ICT solutions, which we have discussed at length in our past work [7]. We touch upon this aspect here briefly for a complete discussion on the feasibility of our solution in the Indian context.

The rest of the paper is organized as follows. Section "Related Work" presents a detailed review of related work and efforts made in ICT, particularly wireless sensor networks (WSN) and remote sensing techniques for precision agriculture. Section "Test Bed Design and Setup" provides a comprehensive description of the setup and design of our experiment and test bed. While Sect. "Implementation" discusses the implementation aspect of our work involving sensors, wireless sensor nodes, and gateway, Sect. "Results and Discussion" discusses the results obtained from our experiments. In the end, Sect. "Conclusion" concludes the paper.

Related Work

Precision agriculture (PA) is a farming concept based on observing and responding to intra-field variations and was initiated with the aim of site specific crop management. PA encompasses components like positioning system, remote sensing, yield mapping, crop and soil sensing, and Analysis and Information Transmission Protocols [14]. A considerable amount of work has been done in PA with various ICT technologies, prominent among them are discussed in this section.

Global Positioning System (GPS), which is a group of 31 satellites, has found enormous applications in user localization. Among its less well-known use cases is to find various topological information about the area. Studies have been carried out using real-time kinematic GPS to know the spatial influence of topological factors on yield of potato [15] by making a high resolution digital elevation model (DEM).

Remote sensing, usually done through aircraft or satellite, is a widely used technique for mapping certain parameters of the soil. The electromagnetic emittance and reflectance data obtained for the crop can provide useful information about the soil type, condition, plant growth, weed infestation, and other parameters. In a prior work, Yang et al. [16] showed that SPOT 5 multispectral imagery in conjunction with spectral angle mapper (SAM), support vector machine (SVM) classification, and maximum likelihood techniques can be used for identifying crop types and estimating crop areas. Another study by Beeri et al. [17] estimated the sugar beet residue nitrogen by alternate satellite sensor imagery and separate spectral models.

Aerial photography method has shown more promising results than satellite imagery due to benefits like operation below cloud and higher spatial resolution. Qin and Zhang [18] have used indices approach to enhance the capability of multispectral remote sensing for disease discrimination at the field level. They used a 5-index image to identify blight disease caused by the pathogen *Phytophthora infestans*. Airborne digital photography (ADP) is becoming famous due to its low cost and high resolution (sub-meter level).

Zhang et al. [19] presented a novel approach by constructing a spectral knowledge base (SKB) of diseased winter wheat plants and mapped the disease severity with the band reflectance of the airborne images obtained from the environment. Although a lot of experiments have been done to improve precision agriculture using remote sensing techniques, there exists a demand to increase the resolution, accuracy, and correlation of remotely sensed data and agronomic parameters.

Crop and soil sensing is an area where extensive amounts of work has been done. Traditionally, plant tissue and soil samples had to be taken to the laboratory for testing, but in recent years, mechanisms have been developed based on direct physical contact and proximate remote sensing technology [20] to perform on-field measurements.

Various soil parameters which are of interest include but are not limited to *soil moisture, conductivity, nutrient analysis, pH,* and *leaf chlorophyll.* Soil sensors of various types such as electrical and electromagnetic sensors, optical and radiometric sensors, mechanical sensors, acoustic sensors, and electrochemical sensors, exist but with certain limitations [21]. Optical sensors with spectral reflectance methods have been used to a fair amount of success to analyze soil moisture, organic content, and some nutrients present in the soil [22, 23]. Idso et al. [24] have shed light on infrared spectrometer technique, which can sense water content in the soil.

Various low-cost and portable water moisture sensors have been developed and are used in developing an automatic irrigation control system. Optoelectronic-based soil organic matter sensor has been developed for both single wavelength and multiple wavelengths. Multiple wavelength sensors require just single calibration and hence showed better flexibility than single wavelength sensors for sensing soil organic matter [25]. Light et al. [26] had developed an automated and multiplexed soil moisture measurement system using granular matrix sensors (GMS). Their system included 144 sensors connected to three multiplexers and eight temperature probes connected to a single multiplexer. They have automated the multiplexed GMS system completely which was accurately predicting changes in soil water content and the occurrence of wetting fronts. Crop and soil sensing seems to be promising for the future, but challenges lay ahead to develop low-cost, portable sensors for in-field measurements of parameters such as soil conductivity, organic content, and nutrients (micro and macro) present in the soil.

Wireless sensor networks (WSN) is an area that is finding increasing applications in precision agriculture. A wireless sensor network is an important component in an end-to-end system that transforms raw measurements to scientifically significant data and results. This end-to-end system includes data collection, calibration, communication, interfacing with databases, web services, data analytics, and visualization tools.

Dursun and Ozden [27] had demonstrated an application of WSN to an area of two acres in Central Anatolia for controlling micro-irrigation of dwarf cherry trees. They have developed a low-cost wireless controlled irrigation system for real-time monitoring of water content of the soil. Their system had three main units: base station unit, solar powered valve unit, and sensor unit for data acquisition.

Musaloiu et al. [28] have developed an experimental WSN for soil monitoring that was installed in Baltimore urban forest initially. Each sensor node measured soil moisture and temperature with a sampling rate of one minute and stored the measurements in local memory. A sensor gateway periodically receives raw measurements and inserts them in a database after calibration. They ran into some challenging technical problems like the need for low-level programming, calibration across space and time, and cross-reference of measurements with external sources.

In India too, precision agriculture using ICT is gaining traction. Pande et al. [29] have proposed a hybrid and energy efficient WSN platform that uses hybrid hexagonal positioning for sensor nodes to achieve better link utilization and solar panels to make it energy efficient. License free, high bandwidth communication, and radio frequency (RF) technologies were used for exchanging multimedia data.

Reddy et al. [30] have developed a personalized agricultural advisory system *eSagu* that enables agricultural experts to deliver their advice to the farmers at regular intervals with the help of digital photographs and other related information without visiting the crop in person. *eSagu* has been used to deliver expert advice to around 6000 farms covering six crops. Results indicate savings in capital cost as well as increase in crop yield.

Sudarshan et al. [31] have developed a comprehensive multimode information and communication system named *Geosense* has been experimented in semi-arid tropics in India under the Indo-Japan initiative. This work combines ICT, GIS, and cloud services. *Geosense* leverages sensor devices with multimode (ZigBee/Wi-Fi/3G/WebGIS) ICTs and integrated location-based services (LBS) to explore high-resolution spatiotemporally distributed sensory data. This is useful in providing and assisting the rural stakeholders with a real-time decision support system (DSS).

Panchard et al. [32] have made an attempt to collect real-time dynamic crop, weather, and environmental parameters using WSN technology to improve agricultural potential. They have studied the feasibility of community-based management through sensor networks (Common-Sense net), in the semi-arid regions of India, that focuses on design and implementation of a sensor network for water management in agriculture with special emphasis on the resource-poor farmers of semi-arid tropical zones (SAT).

We now discuss the details of our setup.

Test Bed Design and Setup

This section presents the description of the experimental setup and various ICT components used in the experiments.

Design of the Experiment

A rectangular field with dimensions of 38 m × 12 m and an area of approximately 5000 square feet was demarcated at IIT Jodhpur's (26.28 N, 73.02 E) old campus, Jodhpur, Rajasthan, India. The area is located neighboring the Thar Desert in India.

There were eight different types of studies carried out on the test bed, description of which are provided in Table 8.1. These studies originated from different combinations arising out of irrigation type (flood/micro), fertilizer used (organic/inorganic) and whether the area under observation used ICT or not. The whole field was divided into 24 equal sized plots, layout of which can be seen in Fig. 8.5.

Two random replications of each study were done in order to account for the differences arising due to temporal and environmental variations, like uneven soil type, tree shade, etc. For example, experiment number 5 was carried out in plots 5a, 5b, and 5c. We have installed a micro-irrigation system in the whole field and

Table 8.1 Plots description

Plot	Type	Irrigation	Fertilizer
1	ICT	Flood	Inorganic
2	ICT	Flood	Organic
3	ICT	Micro	Inorganic
4	ICT	Micro	Organic
5	Simple	Flood	Inorganic
6	Simple	Flood	Organic
7	Simple	Micro	Inorganic
8	Simple	Micro	Organic

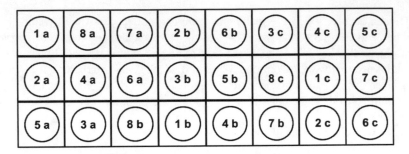

Fig. 8.5 Experimental design

used control valves to emulate situations such as the use of flood irrigation. The description of the various components involved in our setup is provided below.

Micro-irrigation System Setup

Micro-irrigation system was installed on the whole field, layout of which is shown in Fig. 8.6. The description of various components of the system follows thereafter.

1. **Water tank**
     The water tank was kept at 5 feet height from the ground level (Fig. 8.7a). The capacity of the tank used was 300 L A water level indicator in the tank gave

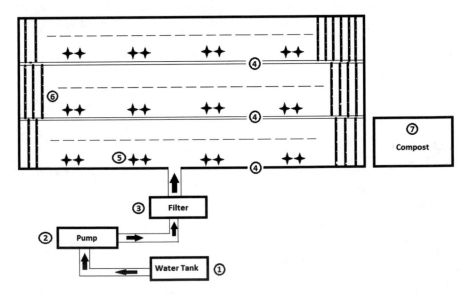

Fig. 8.6 Micro-irrigation test bed layout

a. Water Tank   b. Water Pump

Fig. 8.7 Water tank and pump

us the current level of water. There were one incoming and two outgoing valves linked to the tank. The incoming valve (connected to the main supply source) is opened if water level in the tank is low, whereas the outgoing valve is connected to the pump.

2. **Water pump**

   This is a pump with a 1 hp single phase motor and is used to pump water in the field (Fig. 8.7b). The pump draws water from the tank and supplies it through the filter.

3. **Filter**

   The water from the pump first arrives at the filter (Fig. 8.8a) before filling the main lines. The filter is used to separate coarse materials from water so that it does not choke the emitters or the laterals. Filter is the most important component of the micro-irrigation system, and its design varies from manufacturer to manufacturer. Generally, screen, gravel, or graded sand filters are used with micro-irrigation systems. A combination of more than one filter can also be used depending on the quality of water. Filters must be cleaned and serviced regularly and pressure loss through the filter should be monitored at regular intervals of time. There is a flush valve attached with the filter to flush out coarse materials and other particulate matter. In our test bed setup, we are using a plastic screen filter with a 2 inch opening and a supply rate of 25 m³/h.

4. **Main lines**

   There are 3 main lines running across the length of the field at nearly equal spacing along the breadth. Every main line has 2 PVC pipes. Control valves are connected to one pipe and the second pipe is connected to similar pipes in

a. Filter   b. Control valve   c. Lateral

Fig. 8.8 Filter, valve, and lateral

other main lines. The diameter of each pipe is 63 mm with emitters placed at approximately every 80 cm.

5. **Control valves**

    There are 24 control valves (Fig. 8.8b) spread evenly throughout the field. Every control valve controls one plot, except the last six valves, which controls five rows.

6. **Laterals**

    The laterals (Fig. 8.8c) are flexible PVC pipes, 16 mm in diameter originating from the main lines. There are 46 laterals attached to one set of main line, totalling to 138 laterals along the whole field. The laterals have emitters at 30 cm spacing, out of which water flows at a fixed rate. In our test bed, the emitters discharge water at a constant rate at around 2 L/h.

7. **Compost pit**

    We have tried to setup a non-reactor agitated solids compost bed (process Bangalore). Any kind of plant organic matter is dumped into this pit. This serves the dual purpose of making organic fertilizer while saving at the same time, the organic matter from the traditional practice of burning which is responsible for a considerable amount of environmental pollution. During the rainy season, as water comes in contact with the dumped organic matter in the pit, the matter degenerates faster to become a rich fertilizer that can be used for the next crop.

Implementation

As discussed in Sect. "Test Bed Design and Setup", micro-irrigation was setup in the whole field with 24 control valves in place, one for each plot. Once the approval of soil testing laboratory Jodhpur was received, we chose to plant okra throughout the field. The choice of okra as the experimental crop was influenced by the prevailing weather conditions and short life span of the crop. There were a total of 138 beds, 46 in each line. Around 700 hybrid okra seeds No. 64 were planted at a distance of

Fig. 8.9 Experimental field setup

60 cm from each other. This inter-plant distance was chosen carefully to avoid any root zone interference. The experimental field setup with our WSN setup is shown in Fig. 8.9. We now discuss the system components used for our WSN setup.

System Components

1. **Wireless Sensor Network**

    We have created a wireless sensor network (WSN) which consists of a gateway or base station and sensor nodes. There are six sensor nodes installed in the field. Each node has two soil moisture sensors attached to it which makes a total of 12 soil moisture sensors in plots 1, 2, 3, 4 (and their two replications). It is to be noted that the other plots are not using ICT. The transmission range of the nodes is limited, and the nodes at the far end of the field cannot transmit data directly to the gateway. Hence, a router node or mesh node is also present which transmits the data of the nodes in its range to the gateway. The placement of the router node is such that it is comfortably in the transmission range of every node. The communication protocol used in this wireless sensor network is ZigBee (Fig. 8.10), which operates at 2.4 GHz and is a license-free band. The gateway and sensor nodes used in the setup are described below.

Wireless Sensor Node

The sensor nodes used in the test bed are NI WSN-3202 (Fig. 8.10). It is a four-channel, low power, wireless voltage input device that works with a NI compatible gateway to form a wireless sensor network [34]. These nodes have four analog and four digital channels. The nodes can be operated in two configurations: 'end node' mode and 'router mode.' The six nodes deployed in the field run in 'end node' mode.

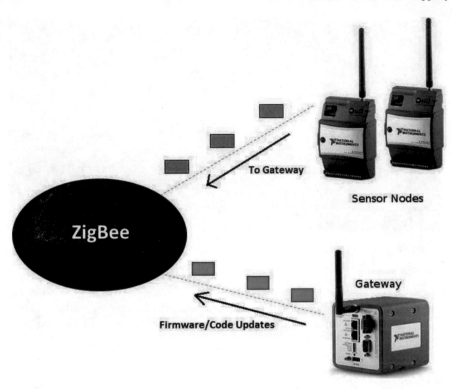

Fig. 8.10 Nodes and gateway communicate through ZigBee [33]

In this mode, the nodes send sensor data received at the analog/digital channels to the gateway. After sending the data, node goes to sleep in order to conserve battery power. In the 'router mode,' the node does not sense through any channel. It just acts as an agent in multihopping data packets all the way to the gateway. It never goes to sleep, and hence, it is recommended to keep it running on mains power supply. The modes of the node can be configured using NI LabVIEW and WSN 1.2 driver software.

We have attached two soil moisture sensors to two analog channels of each node. Although the supported analog channels are four, we have used only two due to power limitations of the node itself. We are powering each node from 4 AA alkaline batteries. In this mode, one terminal of the node acts as a voltage source supplying 12 V and current up to 20 mA. It was observed that only two soil moisture sensors could be supported by the node. If more than two sensors were attached, then the voltage dropped around the terminal, and hence, the readings were not consistent. We have configured the sampling rate of the nodes to 1 min and the activating time of the power terminal to 500 ms just before sampling using NI WSN Pioneer Module. The nodes were enclosed in an NI WSN-3294 Outdoor IP Enclosure (Fig. 8.11a) in order to protect them from wear and tear arising due to harsh weather conditions and the relative ease of clamping and mounting the nodes.

Fig. 8.11 WSN enclosure
and soil moisture sensor

a. Sensor Node   b. Sensor (VH400)

Gateway

The gateway used in this test bed is NI 9792 (Fig. 8.10). This gateway combines
with NI WSN 32xx nodes to form a wireless sensor network [8]. There is a real-
time operating system running in the gateway named VxWorks; hence, the NI WSN
real-time module is needed to use the real-time features. The gateway, on receiving
the data from the nodes, is capable of directly publishing on the web, uploading
to a FTP server, performing HTTP Post and logging data internally or on a USB
stick. We have made a LabVIEW code which on receiving the data converts them
into meaningful information and then uploads it to a database server via HTTP Post.
There is an ethernet port on the gateway through which the visibility of the device is
established on the local area network (LAN). The gateway is kept on mains power
supply as it has to be always active in order to listen to the nodes and process their
data using the LabVIEW code running within it.

2. **Sensors**

  Three kinds of sensors have been used in the test bed, description of which
  are provided here.

Soil Moisture Sensor

Selection of appropriate soil moisture sensors was a challenging task. There are
several kinds of soil moisture sensors available in the market varying from high
cost, high precision to low cost, low precision. We chose VH400 (Fig. 8.11b) from
Vegetronix due to its moderate cost and higher sensitivity. VH400 series soil moisture
probes are small, rugged, high frequency, and less power consuming (as compared to
other similar probes available in the market). Since the probe measures the dielectric
constant of the soil using transmission line techniques, it is insensitive to salinity of

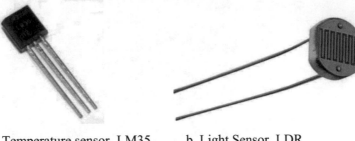

a. Temperature sensor, LM35   b. Light Sensor, LDR

Fig. 8.12 Temperature and light sensors

water and hence will not corrode over time. Also, these probes have a rapid response time of less than 1 sec [35]. This sensor probe can work with a range of input voltages all the way from 3.3 to 20 V. The sensor node supplies 12 V from its terminal, which is comfortably within this range.

The output obtained is analog voltage value between 0 and 3 V. The output has to be mapped to a piecewise linear equation curve to find the value of volumetric water content (VWC) or soil moisture percent in the soil [35].

Temperature Sensor

Sensing temperature is important for an agricultural setup as it has a direct correlation with evapotranspiration and soil moisture estimation. Although we are not using it directly, we store it for the sake of future work. We have used just one temperature sensor (Fig. 8.12a) in the test bed because temperature is a property that varies little over a large area. LM35 series precision centigrade temperature sensor manufactured by Texas Instruments was used because its output voltage is linearly proportional to the Celsius temperature. Also, the sensor is low cost and accurate to a fairly high degree. It is a voltage-based analog sensor with operating range of 4–30 V, making it apt to function with our sensor nodes [36].

Light Sensor

Light dependent resistor (LDR Fig. 8.12b) is a photoresistor whose resistance decreases with the increase in intensity of light. As the intensity of light increases, its resistance decreases, and hence, voltage across it increases (Ohm's Law). By measuring the potential difference across the LDR, we can predict to a fairly good degree the intensity of light falling on the LDR.

3. **Irrigation Automation**

The basic requirement of any irrigation system is to switch on and switch off the pump. For accomplishing these simple tasks, farmers have to sometimes wait to start irrigation at the recommended schedules. A lot of time is wasted for such simple tasks, which could have otherwise been used fruitfully if it could be automated, which we pursue using Information and Communication Technologies (ICT). While one form of automation is to control the water pump based on the moisture content of the soil, another is to use a smartphone to do so. It would be convenient for the farmer to use SMS messages or an application to do so, as s/he might not be in the proximity of the field every time. Not only this, the farmer can also set watering schedules in the application, which would turn on the pump accordingly. We have also tried to prevent misuse of the system by implementing a security mechanism. We have two kinds of users in our system, 'master' and 'slave.' Both master and slave can switch on and off the pump by sending messages S1 and S0, respectively. Any message sent by user other than master or slave will be ignored by the system. The master can change the slave number by sending a mobile number as message prefixed by alphabet 'm.' The number of the master is fixed and cannot be changed by sending any SMS message. It can only be changed by re-burning the code on the microcontroller, which would require advanced knowledge of embedded systems. The 'master' here is analogous to 'administrator' in computer systems. The description of the components used in automating the irrigation system is given in this section.

Microcontroller, GSM Shield, and Relay

Arduino Mega 2560 is an open hardware microcontroller board based on ATMega2560 microcontroller [37]. The GSM/GPRS shield mounted on the board sends SMS messages through serial ports. Our algorithm on Arduino receives the message from the GSM shield, decodes, and authenticates it before further processing. The master and slave numbers have been written on Arduino's EEPROM, which ensures that the numbers are not wiped off when Arduino is powered off. On receiving the message to switch on the pump (S1), the 12th digital pin is driven high. This switches the relay coil attached to this pin and starts the pump. Similarly, on receiving the message to switch off the pump (S0), 12th pin is driven low and relay coil switches back to its previous position.

We have used SIM900-based GSM/GPRS shield for our setup. It is mounted on top of the Arduino board (Fig. 8.13) and has both GSM and GPRS capabilities. It is operated using AT commands and can work at multiple baud rates. We have set a baud rate of 9600 bps and the shield communicates with the Arduino using serial ports.

Once the microcontroller receives the message to turn on/off the pump, it has to actually perform the task. For this, it uses a relay, which is an electromagnetic switch and has the capability to switch circuits when the microcontroller pin is driven high or low. For this work, we have used a 5 V relay as shown in Fig. 8.13. We have connected the relay as per the configuration shown in Fig. 8.14, which is essential to prevent it from damage due to large currents.

4. **Database Server and Web Portal**

     We have dedicated a separate server in our laboratory for collecting the test bed data. All the data collected by the gateway is posted to the server via HTTP Post requests. We have designed a simple web portal that enables one to visualize

Fig. 8.13 GSM shield mounted on Arduino. Relay board is attached to the Arduino

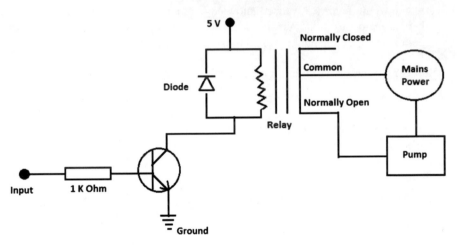

Fig. 8.14 Relay board circuit with transistor and diode

the soil moisture data of the plots in a graphical form (Fig. 8.15), upon specifying the time and date ranges of the data which s/he wants to view.

5. **Android Application**

We have designed an Android application keeping the farmer as the end user in focus. The application has two options, 'Field Data' and 'Remote Control.' On pressing the 'Field Data' button, the user is presented with data from different sensors. The 'Remote Control' button on the main screen can be used to switch on and off the water pump. There is also an option of setting the system in 'Smart Mode.' In this mode, manual participation is totally removed, and the system takes irrigation decisions based on predefined thresholds as discussed in the next section.

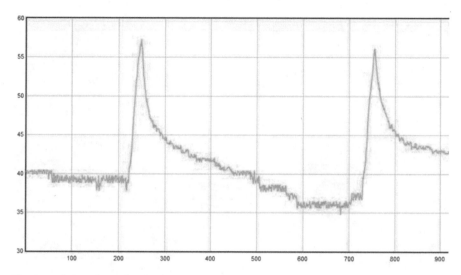

Fig. 8.15 Soil moisture variation over a day. X: time, Y: soil moisture percent

System Description

After looking into the components of the system in bits and pieces, this section gives a holistic view of the system developed by us at IIT Jodhpur, which we call the Agricultural Information and Automation System (AIAS). As seen from Fig. 8.16, the sensor nodes are spread in the 5000 sq. feet or 456 sq m plot. Sensors are attached to the analog input channels of the battery powered nodes. It was observed that at a sampling rate of 1 min, the battery in the node lasts for around 18 days. The gateway, which is always on, receives the data sent from the nodes using ZigBee communication. The LabVIEW code running in the gateway accesses the values received, processes it to a meaningful form, and then uploads the data to the server by doing an HTTP Post. The processed values are stored in the database along with the time stamp. Now the data received from the sensors is used to make decisions for water application in the field. We have defined two thresholds, namely upper threshold and lower threshold. There is a scheduled script running on the server that looks at the average value of the last five samples, and if it is below the lower threshold, then the user gets an email alert to switch on the pump. Similarly, if the average exceeds the upper threshold, the user will get the alert to switch off the pump.

The web portal developed by us uses the data stored in the database to generate plots with the help of an open source library named Flot [38]. The user has to first register to the web portal and then login with his/her credentials to access the portal. Plot name, date range, and time range has to be provided in order to visualize the data. The Android-based smartphone application also uses the database to display sensory data and to make irrigation decisions.

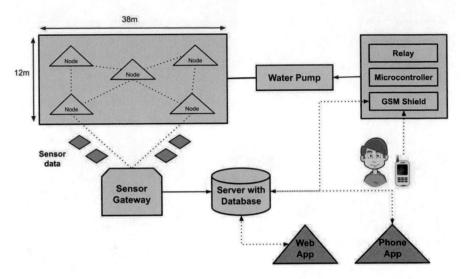

Fig. 8.16 AIAS—macro-level system view

Results and Discussion

The okra seeds were planted at the onset of the summer season. The first set of yield of okra fruit was obtained on the 40th day. The yield obtained and water applied were monitored continuously for 70 days from the day of plantation. It was observed that the number of plants reduced at some stages of the experiment due to harsh climatic conditions of the region, fungus attack, etc. Hence, careful plant count was undertaken at regular intervals.

To reiterate, we have eight categories of plots, description of which can be found in Table 8.1. To compare between the yields obtained from different plots, it is important to define the yield by a common metric that takes into account the reduction of plants in each plot. For this, we devised a metric named *yield ratio* which is defined as the yield obtained (okra produce) for a particular plot divided by the number of plants in that plot (as per the last count). This is essentially the yield per plant per day, under the assumption that the number of plants does not change between two successive counts. Generally, yield was collected after 2 days. But sometimes due to unavoidable circumstances, it was delayed. This delay has been accounted for by the *yield ratio*.

Inorganic and organic fertigation inputs were applied to each plot. The *yield ratios* obtained in different plots are shown in contrast with varying inputs, irrigation, and ICT. The total water applied per plant to the various plots during the time span of the experiment is given in Table 8.2. The rate of water discharge through an emitter in the micro-irrigation system is assumed to be 2 L/h as specified by the manufacturer.

It could be observed from Fig. 8.17a, b that micro-irrigation with ICT is having a *yield ratio* higher than flood irrigation with ICT for most of the samples. In Fig. 8.18a, b, it can be seen that micro-irrigation performs better than flood irrigation for most

Table 8.2 Amount of water applied per plant

Plot	Water (Liters)	Type
1	138	ICT, flood, and inorganic
2	138	ICT, flood, and organic
3	105	ICT, micro, and inorganic
4	105	ICT, micro, and organic
5	161	Simple, flood, and inorganic
6	161	Simple, flood, and organic
7	128	Simple, micro, and inorganic
8	128	Simple, micro, organic

of the time. However, in Fig. 8.19a, b, a trend reversal is seen, where simple flood irrigation has performed better than flood irrigation with ICT. This clearly shows that the threshold set for flood irrigation was lower than optimal, and hence, plants starved in these plots. Figure 8.20a, b show again that micro-irrigation with ICT has outperformed simple micro-irrigation. In Fig. 8.21a, the trends are not clear, but it can be seen that for most of the time, ICT with micro-irrigation has performed better than other types of plots. Also, the worst performer here is ICT with flood due to incorrect estimates of its water requirements. Figure 8.21b shows ICT with micro-irrigation as the clear winner among all others, achieving a *yield ratio* of around 1.6, which is the highest among all the variants. No significant differences in *yield ratios* for plots with organic and inorganic fertigation are recorded.

It can be observed from Table 8.2 that micro-irrigation accompanied with ICT clearly consumes less water than all the other methods. The maximum water was consumed by the plots irrigated with flood irrigation. Water consumption patterns in decreasing order are as follows:

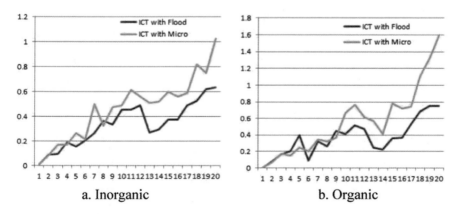

a. Inorganic   b. Organic

Fig. 8.17 Comparison between ICT with flood and ICT with micro-irrigation. X: days, Y: yield ratio

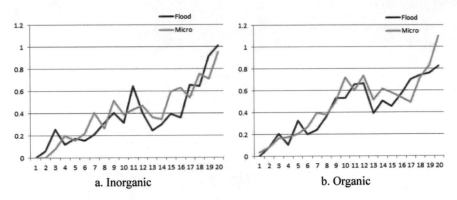

a. Inorganic   b. Organic

Fig. 8.18 Comparison between simple flood and simple micro-irrigation. X: days, Y: yield ratio

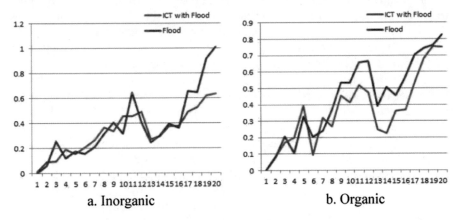

a. Inorganic   b. Organic

Fig. 8.19 Comparison between ICT with flood and simple flood irrigation. X: days, Y: yield ratio

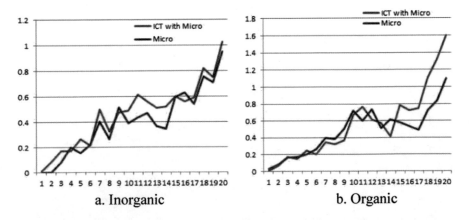

a. Inorganic   b. Organic

Fig. 8.20 Comparison between ICT with micro and simple micro-irrigation. X: days, Y: yield ratio

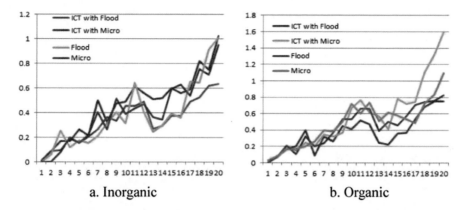

a. Inorganic   b. Organic

Fig. 8.21 Comparison between ICT–flood, ICT–micro, simple Flood and simple micro-irrigation. X: days, Y: yield ratio

$$\text{Flood irrigation} > \text{Flood irrigation} + \text{ICT}$$
$$> \text{Micro} - \text{irrigation} > \text{Micro} - \text{irrigation} + \text{ICT}$$

It is a well-known fact that in most parts of India, flood irrigation is practiced, mainly due to two reasons. First, India is among the top producers of water intensive crops such as rice and sugarcane. Second, there is a lack of awareness as far as micro-irrigation is concerned. Table 8.2 clearly indicates that by using micro-irrigation with ICT, maximum water savings of around 56 L per plant was achieved in 70 days. This translates to a water saving of 35% when compared with flood irrigation. Assuming a crop season of 120 days, say, we have 500 plants in a farm. The total water saving which can be achieved is around 48,000 L, which by no means is a small amount. This is a significant finding of this work.

For micro-irrigation, Plappally et al. [6] have shown that the amount of water applied can be converted into equivalent energy. As observed from Table 8.2, the amount of water applied per crop to micro-irrigated plots is $\sim$ 128 L in 70 days. The total area of the field is 456 m^2, out of which one-fourth, i.e., 114 m^2 belongs to simple micro-irrigation. Assuming an average of 125 plants in these plots, water applied per hectare in one year comes out to be $\sim$ 7218 m^3/ha/yr. Plappally et al. [8] have mentioned 0.22 as the energy intensity of micro-irrigation, using which the equivalent energy consumed can be calculated to be $\sim$ 116 KW h/ha/yr for the okra plant. This energy varies according to the water requirements of crops.

On the one hand, a considerable amount of water saving was recorded with the incorporation of present ICT innovation. The other commercially available ICT solutions are costly, and often out of reach for farmers from developing countries. This can be clearly seen from the first two rows of Table 8.3 presents the cost estimation for a package of WSN solutions from different vendors, having four wireless sensor nodes and one gateway. This table has been reproduced from Adil et al. [7]. It is illustrated that a cheaper alternative to the commercially available (but costly

Table 8.3 Cost comparison of different WSNs [7]

Component	Particulars and quantity	Price (INR)
eKo environmental wireless monitoring system [39]	4 eKo sensor nodes (EN2120), 1 eKo gateway with server (EG2120), 1 eKo base station (EB2120)	331,120.00
NI-WSN [40]	1 NI 9792, integrated WSN gateway 1 NI PS-15 power supply 4 sensor node (3202), 1 LabVIEW WSN Module software	318,500.00
Indigenous wireless sensor devices developed at IIT Jodhpur [7]	4 WS nodes, 1 WS gateway	13,010.00

WSNs) could be readily built. Row 3 of Table 8.3 shows that a working WSN solution could be built at a cost that is around 25 times less than the other commercial options [7]. Adil et al. 2015 also enumerate that the performance in terms of battery lifetime and data acquisition is at par with the available solutions [7].

Conclusion

Sensing and data collection-based control of soil moisture, temperature and electrokinetic soil property solutions have great potential in conserving water and energy in the agricultural domain. This article practically simulates controlled and uncontrolled studies with flood as well as micro-irrigation setup, respectively, in a small arid soil plot near to an area close to Thar Desert, India. The experimental studies enumerated in this work provided savings up to 35% water by volume with the use of ICT, without compromising with the yield. In fact, the yield increased significantly by using ICT, which can be attributed to the irrigation on demand (lOD) strategy used. A low-cost web as well as smartphone applications for operating and controlling the irrigation setup is showcased. The application can be operated by a farmer with little knowledge of text-based messaging systems in a multilingual mode. As part of the future work, the app could be augmented with features indicating chemical requirement, nearby grain elevator or crop storage availability, soil analysis, expert help, and agri-related marketing updates. Other directions for future work would be to fabricate and build WSN setup suitable for rugged conditions and resilient outdoor environments.

References

1. Water Scarcity. https://www.unwater.org/water-facts/scarcity/
2. Water in Agriculture. https://www.worldbank.org/en/topic/water-in-agriculture
3. Sector-wise GDP of India. https://statisticstimes.com/economy/country/india-gdp-sectorwise.php

4. Press Information Bureau. https://pib.gov.in/newsite/PrintRelease.aspx?relid=199881
5. Plappally, A.K., Lienhard, J.H.: Energy requirements for water production, treatment, end use, reclamation, and disposal. Renew. Sustain. Energy Rev. **16**(7), 4818–4848 (2012)
6. Plappally, A.K., Lienhard, J.H.: Costs for water supply, treatment, end-use and reclamation. Desalin. Water Treat. 1–33 (2013)
7. Adil, A., Badarla, V., Plappally, A.K., Bhandari, R., Sankhla, P.C.: Development of affordable ICT solutions for water conservation in agriculture. In: 7th International Conference on Communication Systems and Networks (COMSNETS), pp. 1–6. IEEE (2015)
8. Ni wsn gateway 9792. http://www.ni.com/pdf/manuals/372998b.pdf
9. Franmeier, D.D., Elliot, W.J., Workman, S., Huffman, R.L., Schwab, G.O.: Soil and Water Conservation Engineering, Thomson, 5E (2006)
10. Prasad, R.N., Bankar, G.J., Vashishtha, B.B.: Effect of drip irrigation on growth, yield and quality of pomegranate in arid region. Indian J. Hortic. **60**(2), 140–142 (2003)
11. Kuslu, Y., Sahin, U., Kiziloglu, F.M., Memis, S.: Fruit yield and quality, and irrigation water use efficiency of summer squash drip-irrigated with different irrigation quantities in a semi-arid agricultural area. J. Integr. Agric. **13**(11), 2518–2526 (2014)
12. Şimşek, M., Kaçıra, M., Tonkaz, T.: The effects of different drip irrigation regimes on watermelon [Citrullus lanatus (Thunb.)] yield and yield components under semi-arid climatic conditions. Aust. J. Agric. Res. **55**(11), 1149–1157 (2004)
13. Mostafa, H., El-Nady, R., Awad, M., El-Ansary, M.: Drip irrigation management for wheat under clay soil in arid conditions. Ecol. Eng. **121**, 35–43 (2018)
14. Mondal, P., Basu, M., Bhadoria, P.: Critical review of precision agriculture technologies and its scope of adoption in india. Am. J. Exp. Agric. **1**(3), 49–68 (2011)
15. Persson, A., Pilesjo, P., Eklundh, L.: Spatial influence of topographical factors on yield of potato (solanum tuberosum l.) in central Sweden. Precision Agric. **6**(4), 341–357 (2005)
16. Yang, C., Everitt, J.H., Murden, D.: Evaluating high resolution spot 5 satellite imagery for crop identification. Comput. Electron. Agric. **75**(2), 347–354 (2011)
17. Beeri, O., Phillips, R., Carson, P., Liebig, M.: Alternate satellite models for estimation of sugar beet residue nitrogen credit. Agr. Ecosyst. Environ. **107**(1), 21–35 (2005)
18. Qin, Z., Zhang, M.: Detection of rice sheath blight for in-season disease management using multispectral remote sensing. Int. J. Appl. Earth Obs. Geoinf. **7**, 115–128 (2005)
19. Zhang, J., Huang, W., Li, J., Yang, G., Luo, J., Gu, X., Wang, J.: Development, evaluation and application of a spectral knowledge base to detect yellow rust in winter wheat. Precision Agric. **12**(5), 716–731 (2011)
20. Sudduth, K., Hummel, J., Birell, S.: Sensors for site-specific management. Am. Soc. Agric. 183–210 (1997)
21. Adamo, F., Andria, G., Attivissimo, F., Giaquinto, N.: An acoustic method for soil moisture measurement. IEEE Trans. Instrum. Meas. **53**(4), 891–898 (2004)
22. Rossel, R.V., Walvoort, D., McBratney, A., Janik, L., Skjemstad, O.: Visible, near infrared, mid infrared or combined diffuse reflectance spectroscopy for simultaneous assessment of various soil properties. Geoderma **131**, 59–75 (2006)
23. Stamatiadis, S., Christofides, C., Tsadilas, C., Samaras, V., Schepers, J., Francis, D.: Ground-sensor soil reflectance as related to soil properties and crop response in a cotton field. Precision Agric. **6**(4), 399–411 (2005)
24. Idso, S., Jackson, R., Pinter, P.J., Reginato, R., Hatfield, J.: Normalizing the stress-degree-day parameter for environmental variability. Agric. Meteorol. **24**, 45–55 (1981)
25. Hummel, J., Gaultney, L., Sudduth, K.: Soil property sensing for site-specific crop management. Comput. Electron. Agric. **14**, 121–136 (1996)
26. Light, J.E., Mitchell, A.R., Barnum, J., Clinton, C.: Granular matrix sensors for irrigation management (1991)
27. Dursun, M., Ozden, S.: A wireless application of drip irrigation automation supported by soil moisture sensors. Sci. Res. Essays **6**(7), 1573–1582 (2011)
28. Musaloiu, R.E., Terzis, A., Szlavecz, K., Szalay, A., Cogan, J., Gray, J.: Life under your feet: a wireless soil ecology sensor network

29. Pande, M., Choudhari, N., Pathak, S., Mukhopadhyay, D.: H2e2: a hybrid, hexagonal and energy efficient WSN green platform for precision agriculture. In: Proceedings of the 12th International Conference on Hybrid Intelligent Systems (HIS) (2012)
30. Reddy, P., Ramaraju, G., Reddy, G.: esagu: a data warehouse enabled personalized agricultural advisory system. In: Proceeding of the ACM SIGMOD International conference on management of data. ACM (2007)
31. Sudharsan, D., Adinarayana, J., Tripathy, A., Ninomiya, S., Hirafuji, M., Kiura, T., Desai, U., Merchant, S., Reddy, D., Sreenivas, G.: Geosense: A Multimode Information and Communication System. ISRN Sensor Networks (2012)
32. Panchard, J., Rao, S., Prabhakar, T., Hubaux, J.-P., Jamadagni, H.: Commonsense net: a wireless sensor network for resource-poor agriculture in the semiarid areas of developing countries. Inf. Technol. Int. Dev. 4(1), 51–67 (2007)
33. Wsn node 3202 and wsn gateway 9792. http://sine.ni.com
34. Ni wsn 3202. http://www.ni.com/pdf/manuals/372775e.pdf
35. Vegetronix soil moisture sensor. http://www.vegetronix.com/Products/NG400
36. LM35 precision centigrade temp. sensors. http://www.ti.com/lit/ds/symlink/lm35.pdf
37. Atmega2560. http://www.atmel.com/Images/doc2549.pdf
38. Flot. https://www.flotcharts.org/
39. eKo environmental monitoring system. http://www.memsic.com/wireless-sensor-networks/
40. Ni wireless sensor networks. http://www.ni.com/wsn/

Chapter 9
Adoption of Climate Smart Agriculture (CSA) Technologies in Sri Lanka: Scope, Present Status, Problems, Prospects, Policy Issues, and Strategies

G. V. T. V. Weerasooriya and S. Karthigayini

Introduction

Agriculture plays a major role in the economy of Sri Lanka. It is often considered the basis of all civilization. It is a part of everything from the food people eat to the clothing people wear. It shapes many of the traditions and values that this world was built on and it is the science, art and occupation of producing crops, raring livestock, and cultivating the soil [6].

In many nations, agriculture is the primary supply of livelihood, using most of the population and often the most important contributor to Gross Domestic Product [2]. Agriculture faces diverse challenges namely food protection, agricultural practices, land redistribution, natural agriculture, conventional agricultural practices v/s sustainable agricultural practices, screw ups in marketplace structures, exchange limitations, volatile and ineffective socio-financial regulations, lack of facts, accessibility of infrastructural and financial structures, continuously increasing population strain, insufficient assets, inappropriate agronomic practices, weather adjustments, and environmental degradation [7]. As a result, the number one stake holders who are agro-manufacturers, agricultural customers, the government and the surroundings are stricken by these challenges [19]. These challenges are further exacerbated by way of the consequences of climate variability and is specifically weather dependent. Hence, agriculture remains one of the maximum inclined sectors to climate variability and change [13].

Agriculture is the spine of Sri Lanka's rural economic system. It's contributing to approximately 6.9% of the national GDP out of which the fisheries area contributes around 1.3% and the cattle 0.6%. Over 25% of Sri Lankans are hired within the

G. V. T. V. Weerasooriya (✉) · S. Karthigayini
Department of Agricultural Engineering and Soil Science, Faculty of Agriculture,
Rajarata University of Sri Lanka, Anuradhapura, Sri Lanka
e-mail: gvtvw@agri.rjt.ac.lk

© Centre for Science and Technology of the Non-aligned and Other
Developing Countries 2023
K. Pakeerathan (ed.), *Smart Agriculture for Developing Nations*, Advanced Technologies
and Societal Change, https://doi.org/10.1007/978-981-19-8738-0_9

agricultural sector. Even though Sri Lanka is a fertile tropical land with the potential for the cultivation and processing of a selection of plants, issues inclusive of productiveness and profitability bog down the growth of the sector [9]. Sri Lanka's weather is considered tropical monsoonal, and it has three climatic zones together with wet zone (WZ), dry zone (DZ), and intermediate zone (IZ). In line with climate predictions directly affecting the rural activities inside the wet zone and dry zone [32]. Sri Lankan agriculture has already felt the effect of negative climate events and weather changes. In Sri Lanka, subsistence farming systems are the foremost farming structures for several small-scale farmers engaged in conventional agricultural practices which include cut back and burn and transferring cultivation. With population boom, the expansion on arable land will probably continue in lots of areas [32].

Agriculture Sector in Sri Lankan Context

Sri Lanka is one of the South Asian countries which are located underneath the Indian peninsula. The majority of the Sri Lankan population (70%) lives in rural areas wherein farming is widely practiced and one tenth of the population lives beneath the poverty line [31]. Agriculture area plays a vital position in contributing to the USA's economy due to the fact agriculture sector is the main supply of livelihood for the agricultural human beings [33].

There are four primary components in agriculture zone such as cultivation of plants, rearing livestock, fisheries, and forestry [32]. Sri Lankan agriculture zone incorporates both plantation and non-plantation agriculture. Plantation agriculture sector especially includes production of tea, rubber, and coconut. It is an export-oriented sector. Non-plantation agriculture is focused on production of food plants inclusive of paddy, different cereals, pulses, condiments, vegetables, and fruits. It's far specifically for home consumption. Agricultural sector in Sri Lanka is particularly for the fulfillment of food security and improving the incomes of smallholder farmers [11].

Sri Lankan agriculture sector is now moving toward organic agriculture which is healthier. In the agriculture sector, crop cultivation seems to be unprofitable over the time due to adverse impacts on climate changes. In Sri Lankan agriculture sector, there is a need of having training for farmers to encourage cultivation of fruits and vegetables up to the standard like rice production [15].

Livestock sub-sector plays an important role in the agriculture industry to improve the standard of living. It contributes 0.8% to the country's GDP. With a small proportion, many livestock products have to be imported. In Sri Lanka, main livestock products are milk, meat, and eggs [9]. In past years, animal power was used in the cultivation of rice and vegetables but nowadays it has been replaced by modern technologies. However, animal husbandry contributes in the rural economy for improving the living conditions of farmers in the country [11].

Prasannath [26] concluded that other sub-sector of fisheries sector is also very important in the agriculture industry. It performs a key position in Sri Lanka's social and monetary life. Fish merchandise is the foremost supply of animal protein in Sri

Lanka. The fisheries zone of Sri Lanka consists of three main sub-sectors such as coastal, offshore and deep sea, and inland and aquaculture.

In the beyond years, forests performed a key role inside the rural economic system in Sri Lanka. The dependency of communities on wooded area assets for subsistence needs has steadily declined because of industrialization and allied social transformation. The principle reason of deforestation is for alternative land uses. This has located country's rich biodiversity in an everlasting risk. Sri Lanka has added the concept of sustainable development and incorporated it in new woodland sector rules to provide extra safety for ultimate forest resources [32].

At present, agriculture sector is fluctuating due to several reasons such as lack of awareness about cultivating techniques, poor practices, lack of infrastructure facilities, water and land scarcity, urbanization, high cost of production, and adverse weather conditions. Hence, it is necessary to improve the farmers' knowledge, skills, and entrepreneurship in order to attain the productivity [11].

Weather exchange can also have effects on all human beings because of its dangerous threats to the environment and agricultural yields everywhere in this globe. Weather is one of the most crucial factors of distribution and abundance of species [16]. Severe climate situations together with heavy rains, high temperature, and high wind pressures have an awful lot influence on the rural activities. The financial results are already being felt by means of farmers and across international deliver chains. Agricultural agencies pick out weather change as a critical long-time chance in supply management. Weather and agriculture are strongly interrelated prevalent methods, and for this reason changes in weather have an effect on agricultural activities [16].

Therefore, to conquer the troubles, Food and Agriculture Organization (FAO) has recognized that for agriculture to feed the sector in a way which could defend sustainable rural improvement, it has to be as "Climate Smart Agriculture" (CSA), as defined and provided by means of the FAO at the Hague conference on agriculture, food safety, and climate exchange in 2010 [22].

In Sri Lanka, traditional and weather model techniques co-exist [15]. Variation to intense weather events, in particular flooding and drought is most significance. As an end result, farmers have adopted many CSA practices aimed at water control, soil conservation, residue management, land use making plans, cattle control, fodder and agroforestry management, biogas production, and so forth [9]. CSA practices appear not to be uniformly adopted across production systems. There are many factors which influence adaptation of CSA technologies such as household characteristics, economic status, and farm characteristics, and climate variations. The faculties of Sri Lankan universities are accountable for higher training, schooling, and research in CSA strategies as well as provincial departments provide CSA technology associated statistics and education on the community and farm level [24]. Private investment in CSA specifically comes from farmer's very own resources and restrained loan and coverage schemes. Even though Sri Lanka has accessed CSA finance via home and worldwide agencies, the country has limited assist from non-public and financial sector organizations [32].

In developing countries for small holder farmers to adopting CSA technologies than mitigation, the opportunities for greater food security and increased income together with greater resilience will be more important [13]. Climate Smart Agriculture technologies are sustainably support food security, incorporating the need for adaptation and the potential for mitigation into development strategies to enhance the capacity of the agriculture sector. In Sri Lanka due to the adverse effect of climate change, there is a greatest negative impact on poor households as they have a low adaptive capacity to climatic change [25].

Climate Smart Agriculture is not defined as set of practices or a wholly new type of agriculture. As a substitute, it is an approach that integrates specific strategies below a climate trade umbrella. In the face of climate change, it offers farmers tools and a pathway to make their operations and livelihoods greater productive and resilient, as properly as it reduces their climate affects [30]. Some common examples of CSA technologies are drip irrigation, sprinkler irrigation, raised beds, composting, soil management, drainage management, residue management, fertilizer management, gender empowerment, land use planning, cover crops, mulching, livestock management, agro-forestry, fodder management, etc. Suitable technologies will differ due to several factors such as region, ecosystem, climate, crops, farmers' perceptions, and farmer's adaptation behavior [12].

These technologies are crucial to fulfill climatic challenges by way of increasing resilience to climate extremes, adapting to climate exchange, and decreasing agriculture's greenhouse gas (GHG) emissions and additionally to aid sustainable food protection and agricultural productivity [17]. In the coming decades particularly in dry zone areas where adaptive capacity is weaker, climate risks to crops, livestock, and fisheries are expected to increase in a high potential. There are several problems due to changing climate patterns; it will affect people who are engaging in farming activities in all ecosystems. But people who are doing farming in dry zone areas face more acute challenges. Majority of the farm families in dry zone of Sri Lanka depend directly on paddy and other field crop cultivation for their living [32].

In dry zone already stricken by excessive poverty tiers, due to terrible land and water availability therefore, it's hard to escape from the climatic change with negative rainfall, greater common droughts, extreme temperature, and the animal or new crop pests and sicknesses. The increased variation in rainfall and decrease in the overall quantity affect production stage in dry zone [32]. Therefore, applying CSA technologies inside the dry zone goes to require a great deal more interest in the future. So, there is an urgent want to put into effect CSA technology to assist small holder farmers adapt to varying climate [9].

Dry zone consists of the primary percentage of agricultural land in Sri Lanka. Farmers begin their cultivation with the start of Maha season rainfall. The northeast monsoon brings rainfall to dry quarter regions or Sri Lanka.

Paddy cultivation is the important crop grown in the dry area. The call for water sources is anticipated to increase due to the enlargement of place under irrigated agriculture, growing population, urbanization, and industrialization [32]. Prasannath [26] discussed approximately, a few serious implications on agriculture, food safety, health, industries, and energy due to the declining tendencies of the watershed areas. Delayed monsoon rains and a growth inside the frequency of droughts and floods typically affect the quantity of cultivation, harvesting, and yields. The weather adjustments of temperature and rainfall will possibly to have direct influences on soil moisture. Majority of the farm families in dry zone of Sri Lanka depend immediately on paddy and different vegetation cultivation for his or her dwelling.

There are various adverse weather effects happening due to climatic variations. Due to these climatic variations, farmers are faced with many challenges in their livelihood and economic status. Therefore, farmers are adopting CSA technologies with their traditional/conventional agriculture practices in order to overcome the above challenges. CSA technologies sustainably increasing food security by increasing agricultural productivity and incomes, building resilience, and adapting to climate change and reduce greenhouse gas emissions. There are many factors that influence extent of adoption of CSA technologies. It is important to identify these drivers which influence farmer's decision to adopt CSA technologies.

Therefore, this study mainly concerns on how farmer's adoption behavior for CSA technologies are often linked with climatic conditions and what are the CSA technologies mostly adapted by farmers and the scope, present status, problems, prospects, policy issues, and strategies related to the adoption behavior of Climate Smart Agriculture technologies in Sri Lanka while reviewing water smart, energy smart, nutrient smart, carbon smart, weather smart, knowledge smart, yield smart, and soil smart technologies.

Challenges in Agriculture Industry

Agriculture is a major and vital occupation in developing countries because without farming, the world would starve to death. Small and marginal holdings agriculture is important for raising agriculture growth, food security, and to improve the economy. The nation depends on the performance of these small and marginal farmers for the future of sustainable agriculture growth and food security. Especially in developing countries, global food demand is increasing due to population growth [1].

Most of the farmers in globally continue to face splendid demanding situations in everyday lives. Most farmers nowadays are small holder or subsistence farmers who develop plants and rear animals simply to feed themselves and their households, additionally majority lives in rural and suburban communities. There are several short- and long-term demanding situations that face agriculture over the following 50 years [28].

Maya [16] recognized that there are numerous challenges confronted through farmers inside the agriculture sector. These are high price of finance, insufficient physical infrastructure to assist the agriculture zone, negative farming strategies, market place oriented manufacturing and profitability, economic increase, radical monetary transformation, limited opportunity to excellent farm inputs, lack of marketing opportunities, insufficient manufacturing and post-harvest techniques, insufficient disease management facilities, lack of care by farmers on higher farming methods, low profits to buy farm inputs, low value addition in agriculture produce, restrained public agricultural institutions, insufficient storage and high post-harvest losses, susceptible implementation of agricultural legal guidelines and regulations, land issue in a few areas, water shortage, hard work shortage, and unfavorable climatic changes [31].

Climatic change is considered as one of the most challenging current global issues. Climate is the totality of environmental factors such as rainfall, air temperature, solar radiation, air circulation, relative humidity, and their long-term variation. Agriculture sector is highly sensitive to both short- and long-term changes in the climate. Small holder farmers are particularly vulnerable to these changes in the climate that reduce productivity and affect their weather dependent livelihood systems negatively [12].

Overview of Climate Change and Climate Smart Agriculture Technologies

Qui et al. [27] talk about the climate change. It is defined by the change in the climate over time due to variability of natural processes, external forces, and continuous anthropogenic activities in the components of the atmosphere or land use. The global climate changes are happening in an increasing pattern. It is being experienced in various forms including as temperature rise, sea level rise, droughts, floods, hurricanes, and landslides. Climate change is creating serious impacts on human well-being and will continue to expose damages in the future [5]. Agricultural systems such as cultivation of crops, livestock management practices are highly depend on climatic factors and highly affected by climate change and variations.

Developing countries are highly vulnerable to climate change [31]. Most serious impacts of global climate change will be felt on smallholder farmers in the developing countries. Direct impacts of climate change over the next few decades will be felt on agricultural and food systems. It is mainly affecting the agricultural productivity [30].

In Sri Lanka, the main climatic parameters such as rainfall and temperature have a significant impact on agriculture. Effects of rainfall relative to temperature are more important for agriculture in a tropical country like Sri Lanka. High rainfall variability induces extreme rainfall events such as floods and no rainfall causes droughts which are the most prominent climate based natural disasters causing damages to agriculture in Sri Lanka [11].

Human activities will be severely affected by climate change. In Sri Lanka, climate change will affect the agricultural sector due to the variability in rainfall, increase temperature, and elevated CO_2 concentration. Climate change will equally affect both plantation and non-plantation agriculture. However, the impact non-plantation agriculture will be more severe because of the implications of climate changes because non plantation agriculture depends on specific climate conditions. As the majority of smallholder farmers in dry zone of Sri Lanka are dependent on rain for their cultivation, they are the people most vulnerable to variability and reduction of rainfall since it has implications for farm profitability [11]. To overcome aforementioned climate variation problems farmers who are dependent on agriculture, they have to adapt to the climate changes. There are some adaptation strategies such as using scarce water resources more efficiently, mitigation and adaptation, building flood defenses, developing drought and disease tolerant crops varieties, choosing tree species and forestry practices, integrated pest management, improve the policies, developing infrastructures, and adapting CSA technologies [17]. There are some barriers which are limiting their capacity to adapt to the climate changes such as high dependence on natural resources, small farm sizes, low technology and capitalization, poor infrastructure, and institutional support [23]. But, the adaptation strategies are very important to make concern in communities to minimize climate change impacts. Climate change impacts on the agricultural sector in Sri Lanka can be significantly reduced through appropriate adaptation strategies [32].

CSA is an approach for developing agricultural strategies to secure sustainable food security under climate change. It provides the means to help stakeholders from local to national and international levels identify agricultural strategies suitable to their local conditions. Therefore, among the adaptation strategies CSA technologies are the most appropriate strategy for smallholder farmers in developing countries [31].

FAO [8] defined and presented about Climate Smart Agriculture at the Hague Conference on Agriculture, Food security, and Climate change in 2010. The concept of CSA has been developed to address three pillars: food security, adaptation, and mitigation. Climate Smart Agriculture has gained considerable attention, especially in developing countries; these triple objectives focus on improving food security by sustainably increasing productivity and income, adapting to climate change and reducing greenhouse gas emissions [7].

The CSA concept reflects the integrating agricultural development and climate responsiveness. The concept mainly focuses on achieve food security and broader development according to the changing climate and increasing food demand. Increased planning is very important in order to indicate trade-offs and synergies between the three pillars of productivity, adaptation, and mitigation [3].

Middelberg [19] discussed about CSA mainly focuses on identifying the needs of people for food, fuel, timber, and fiber through science-based actions which are contributing to economic development, poverty reduction and food security, which maintain and enhancing the productivity and resilience of both natural and agricultural ecosystem functions. So through these science based actions, building natural capital and reducing the trade-offs mainly involved in meeting these mentioned goals.

Climate Smart Agriculture is an approach which brings together agricultural practices, policies, institutions, and financing in the context of climate change. This approach is designed to identify operationalize sustainable agricultural development obviously integrating climate change as a major parameter [25]. The Climate Smart Agriculture includes several field-based sustainable management practices based on land agriculture (cropping), livestock management, fisheries, and forestry sectors. Under these primary practices, there are many technologies were included [17].

To control or resistant to climate variations, an individual farmer might adopt climate smart practices by switching vulnerable activities. As an example, farmer can change from crops to heat tolerant animals to overcome adverse weather conditions [30]. If Climate Smart Agriculture to become a success and a reality, an integrated responsive to specific local conditions are required. Integrated landscape approaches and coordination across agricultural sector are essential to capitalize on potential synergies, reduce trade-offs, and optimize the use of natural resources and ecosystem services. Across the productive landscapes, there are several challenges in the environmental, social, and economic dimensions for these challenges CSA technologies coordinate the priorities of multiple countries and stakeholders in order to achieve more efficient, effective, and equitable food systems. While the concept is new and still evolving, many of the practices that make up Climate Smart Agriculture technologies already exist in worldwide and are currently used by farmers to cope with various production risks [6].

Testing and applying different practices are important to expand the evidence base, identifying which practices and extension methods are suitable in each context and identify the synergies and trade-offs between food security, adaptation, and mitigation. CSA technologies and practices present opportunities for addressing climate change challenges, as well as for economic growth and development of agriculture sectors. Climate smartness is ranked in all key production systems in the country, exposing ongoing and potentially applicable practices, as well as practices of high interest for further investigation or scaling out [29].

Amin et al. [2] argued CSA has an important gender dimension. Gender is the basis to understand different social, economic, and cultural roles of men, women, boys and girls in societies. To identify these gender relations, a better understanding can be gained on the gender differences in access to CSA interventions and opportunities. Both men and women are likely wanting to adopt CSA, since these technologies can bring income, increase food availability for the household, support to deal with climatic impacts, training opportunities, and differential access to production resources. However, factors as labor, knowledge, and property of land differ among men and women [22].

In Sri Lanka, CSA technologies are not adopted completely by farmers and still are evolving. Access to new, climate adapted technologies are ensured by the Department of Agriculture (DA), which implements several programs that target farmers in different agro-ecological regions [9]. Technologies and practices which are under water smart, energy smart, nutrient smart, carbon smart, weather smart, knowledge smart, yield smart, and soil smart are key Climate Smart Agriculture practices adopted in by smallholder farmers Sri Lanka. Adoption of CSA practices requires institutional support, especially for smallholder farmers. In particular, medium- and long-range seasonal climate forecasts, better intra- and inter-institutional coordination, and improved market access by smallholders are prerequisites for increased CSA adoption in Sri Lankan agricultural systems [32].

From the understandings of CSA above, it is clear that CSA is primarily concerned with adapting to a changing environment and reducing agricultural losses.

Adaptation Behavior of CSA Technologies

Adaptation strategies which provide several benefits and services on agricultural production systems and rural livelihoods should enhance the health and functional ecosystems [18]. There are some concepts which are used to understand and approach climate change adaptation behavior such as vulnerability, adaptive capacity, resilience, exposure, and climate change adaptation itself. Adaptation measures that are available for farmers to implement CSA include conservation agriculture (CA), sustainable agriculture (SA), organic agriculture (OA), and use of improved high yielding resistant varieties that can endure so many stresses in the field [14].

Climate change adaptation strategies can be incremental or transformational. Incremental actions aim to maintain the essence and integrity of a system or process at a given scale. Incremental changes enhance the response to changing conditions little at a particular time and can be made iteratively [4]. Transformational actions change the basic attributes of a system in response to climate and its effects. Transformational changes are shifting the system into another system. However, both types of adaptation strategies require for different measures for integration, policy planning and resource mobilization [16].

Parry [24] discussed about climate change adaptation options can be categorized by some groups such as physical, social, and institutional. Physical options are used on the ground or landscape level, for example, improved technologies, irrigation infrastructure, and resources management. Social adaptation options are used to reduce vulnerability of social systems and populations, for example, financial services, education, and social safeguards. Institutional adaptation options indicate the institutional, government, and policy context, for example, improving the laws and regulations to enhance the agriculture sector, institutional capacity development and establish policies and programs.

There are two main approaches under the climate change adaptation behavior such as community-based adaptation approach and ecosystem-based adaptation approach. The community-based adaptation approach is used to implementing adaptation actions through assessments of impacts and vulnerabilities. This approach is utilized by relevant stakeholders. It's important to assist communities from the influences of weather change. The ecosystem-based adaptation approach is used to building and sustaining wholesome and useful ecosystems. It is presenting a spread of blessings and services for agricultural manufacturing structures and rural livelihoods. The vulnerability of equipment depends upon its sensitivity to weather adjustments and on its capacity to control those climatic adjustments [12].

An Overview of Adoption Behavior of CSA Technologies in Sri Lankan Context

The need for these technologies which helps to limit weather exchange influences is seen as important. According to the literature review, climate alternate impacts on the rural region in Sri Lanka may be exponentially decreased via appropriate variation strategies [26]. Sri Lankan authorities having clear idea on the significance climate change, has taken several projects in order to address the impacts of climate change [26].

In Sri Lanka because of the generation improvement, the most common technique is to broaden crop sorts to increase tolerance to moisture stress, salinity, and excessive temperature and accelerated CO_2 levels. In Sri Lanka, crop management that includes technology such as planting dates, shortening of the growing season, converting crop varieties, farm animals' control, fodder and forestry control and crop rotation are the maximum time consumed variation technologies amongs farmers [11]. Studies in Sri Lanka have encouraged numerous versions of CSA technologies to limit the effect from weather change. For the unfavorable weather modifications, changing planting time to appropriate rainfall variability, creation of micro irrigation and reduction of irrigation intensity are a number of the proposed strategies [32].

In the dry zone of Sri Lanka, suitable adaptation strategies for climate change impacts on North east monsoon could be overcome by considering a short duration paddy variety to avoid expected decrease in rainfall during the months of January and February. Farmers in dry zone mainly focus on using water saving methods such as drip irrigation with agro wells for other field crop production and cultivation of low water requiring crops such as maize and grain legumes which would be beneficial [9]. For the sustainability of the dry zone agriculture policy makers may consider providing more water to the farming activities. Scientist may consider methods to reduce irrigation water requirement and suitable crops varieties to cope with temperature and water stress. In addition, improve and restore the storage capacity of the Mahaweli reservoirs by desilting them to provide more irrigation water [18].

In 2005, countrywide rainwater harvesting policy turned into formulated to promote rainwater harvesting in the dry zone areas to confirm the predicted decline in rainfall [31]. The village tank adapt by means of farming activities with the identified seasonal decline of rainfall and managing rainwater harvested within them. Paddy cultivation zero tillage situations have enabled a reduction in value of production and better conservation of water without considerably affecting the yield [9].

Factors Influence on Adoption Behavior of CSA Technologies

Farmers' choice is the main factor which influence on adopting CSA technology to overcome climatic impacts. There are some characteristic features of effective efforts which encompass community participation, use of indigenous knowledge and practices, development of technology and combination of conventional technological innovation manner with the information of farmers [28]. If a planner take decision about adaptation strategies, there are two key concerns arise related to decisions such as the profitability for individual farmers and the duration of adoption strategies which provides public benefits [22].

There are some key factors influence on adaptation behavior such as farmers' household characteristics, farmers income status, farm characteristics and climatic variations as well as labors to carry out agronomic practices, information suitable for fodders and initial capital, available space in small land holdings, availability of planting materials, and availability of alternative methods also influence on adaptation of CSA technologies [13]. Model of new technologies, improvements or practices, allocation of various ways of capital could beautify performance essential for the adoption of interventions to acquire the preferred impact within the farming machine [20]. There are some attributes of farmers that directly influence on adaptation measures such as land ownership, type of cultivation method, and land distance with water resources. The factors that influence farmers' adaptation behavior according to the farmers' internal factors such as characteristics, knowledge, perceptions and external forces such as economy, policy, environmental and technology [23]. In addition influence on adaptation behavior of farmers such as household educational status, farming experience, land size, perception of climate change impacts, knowledge and skills, stability of farm productivity, soil protection strategies, income, access to the information, access to the market, extension services, crop diversity, livestock production, institutional, policy, and technological support are found to be the factors influencing the adaptation behavior of CSA technologies [19].

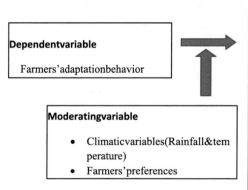

Scope and Status of CSA Technologies in Sri Lanka

The scope and status of sustainable agricultural development are different for different countries. New scope for CSA technologies has aroused due to speedy socio financial changes in Sri Lanka and it has all started to undertake some CSA technologies, mainly on farm related studies. However, the adoption remains very confined. Some representative examples regarding CSA technology in Sri Lanka namely application of CSA for cash crops, plantation crops and small-scale farms, etc.; therefore, exhaustive review on the status of CSA technology adoption is very important.

Adoption and success of CSA technologies in Sri Lanka will depend on the proper design of those strategies. A well-planned number of experiments and analysis are required before application of CSA technologies. With an intention of making awareness among people, first need to involve in uniform crop and soil control, development of manpower and institution for CSA technology, popularization of CSA idea by using mass media communication, seminar, workshop, and so on., whereas in addition should follow random sampling experiments inside regions.

CSA has created scope of remodeling the conventional agriculture, through the way of proper equipment usage and management, to an environmentally friendly sustainable agriculture. Genetic algorithms, wavelength techniques, selection tree, smart microprocessors, genetically engineered plant, biosensors along with other improvement will make CSA to ruin the space not only suitable for developed countries however for developing countries like Sri Lanka as well.

Problems and Prospects in Adoption of CSA Technologies

The main among various constraints in adopting CSA technologies is the mindset of the farmers [10]. It was experienced that convincing farmers is a big hurdle in promoting this technology in agriculture. Further, there are some more problems for adopting namely, environmental problems, lack of awareness about these technologies to farmers, requirement of skilled manpower for operation, etc.

Awareness and adoption change of CSA technologies are stricken by many elements consisting of traits of the farms; land size, land kind, diversification, experience, schooling, hard work days, personality and own family structure of the farmer, family type, size, social interplay, monetary fame; supply of earnings, government support, etc.

Moreover, animals, farm working, and equipment can easily cause incidents by means of accessing farming environments. Additionally, smart structures include heterogeneous gadgets and software from specific manufacturers. Those unique functions may make several protection breaches.

   The promoting of CSA technologies below Sri Lankan context has the prospects specifically, multiplied yields, environmental advantages, decreased cost of manufacturing, reduction of weed prevalence, water and nutrients saving, crop diversification, aid development, etc.

Policy Issues and Strategies for Adoption of CSA Technologies in Sri Lanka

There is a need for policy analysis to understand how these technologies interrelated with other technologies because the Climate Smart Agriculture shows an exponentially increase of change from traditional agriculture. Therefore, in order to promote CSA technology, there are some important policy considerations such as, support for the adaptation of CSA technologies in local environment, support the development of technology, promote payments to environmental service staffs, building relationship between policy makers and farmers, providing loans or subsidy to buy technological equipment, provision of adequate requirements to the farmers, etc.

Conclusion

The review of this study may help policy makers understand how Climate Smart Agriculture technologies perceived by the agricultural producers and how extent these technologies may be used to enhance yield, production, and environmental quality and also provide awareness on problems and solutions in adoption of CSA technologies. Further, it helps to market these technologies to agricultural producers by agro-dealers, extension officers, etc. And also, Climate Smart Agriculture technologies manufacturers and sellers can use this review to identify the demand of their product and services in the future.

Recommendation

This review was done to know about the presence of Climate Smart Agriculture in Sri Lanka. It is recommended to study further about the criteria like scope, status, problems, prospects, policy issues, and strategies in adopting new smart technologies in other countries in order to promote the production.

References

1. Alston, J.M., Pardey, P.G.: Agriculture in the global economy. J. Econ. Perspect. **28**(1), 121–146 (2014)
2. Amin, A., Mubeen, M., Hafiz, M.H., Nasim, W.: Climate smart agriculture: an approach for sustainable food security. Agric.l Res. Community **2**(3), 13–21 (2015)
3. Aryal, J.P., Rahut, D.B., Maharjan, S., Erenstein, O.: Factors affecting the adoption of multiple climate-smart agricultural practices in the Indo-Gangetic Plains of India. Nat. Res. Forum **42**, 141–158 (2018)
4. Asai, K.: Questionnaire survey on farming adaptation for climate variability in Serang Municipality, Indonesia. Mem. Fac. Eng. Yamaguchui **67**(2), 99–106 (2014)
5. Belay, A., Recha, J.W., Woldeamanuel, T., Morton, J.F.: Smallholder farmers' adaptation to climate change and determinants of their adaptation decisions in the Central Rift Valley of Ethiopia. Agric. Food Secur. 1–13 (2017). https://doi.org/10.1186/s40066-017-0100-1
6. Campbell, B.M.: Climate-smart agriculture. Rural **21**(4), 14–16 (2017)
7. Dongre, A.: Agricultural sector: status, challenges and its role in Indian economy. J. Commer. Manage. Thought (2018)
8. Food and Agriculture Organization: Success stories on climate-smart agriculture (2010). https://www.fao.org/climatechange/climatesmartclimate-change@fao.org
9. Gunawardana, A.: Agriculture sector performance in the Sri Lankan Economy: a systematic review and a Metadata analysis from year 2012–2016. In: Agriculture Sector Performance in the Sri Lankan economy, British School of Commerce, Colombo Dissertation for MBA Executives (2018). https://doi.org/10.13140/RG.2.2.34177.92003
10. Hobbs, P.R., Govaerts, B.: How conservation agriculture can contribute to buffering climate change. In: Reynolds, M.P. (ed.) Climate Change and Crop Production, pp. 177–199. CAB International (2010)
11. Jayasinghe-Mudalige, U.: Role of food and agriculture sector in economic development of Sri Lanka: do we stand right in the process of structural transformation. J. Food Agric. **1**(1) (2010)
12. Khatri-Chhetri, A., Aggarwal, P.K., Joshi, P.K., Vyas, S.: Farmers 'prioritization of climate-smart agriculture (CSA) technologies. Agric. Syst. **151**, 184–191 (2017). https://doi.org/10.1016/j.agsy.2016.10.005
13. Lopez-Ridaura, S., Frelat, R., Wijk, M.T.V., Valbuena, D., Krupnik, T.J., Jat, M.L.: Climate smart agriculture, farm household typologies and food security An ex-ante assessment from Eastern India. Agric. Syst. **159**, 57–68 (2018)
14. Manda, L.T.: Farm-level assessment of climate smart agriculture (CSA) indicators in Lushoto district, Tanzania. Thesis report. Farming System Ecology (FSE) Wageningen University and Research Centre (2009)
15. Marambe, B., Pushpakumara, G., Silva, P.: Sustaining food security for sustainable development : setting up research priorities two goals of our time—to achieve sustainable development. In: Developing a Science Technology and Innovation (STI) Agenda for Sri Lanka towards Future Earth. Faculty of Agriculture, University of Peradeniya Department of Agriculture, Peradeniya (2012)
16. Maya, W.: Climate Smart Agriculture for Smallholder Farmers in Southern Africa Climate Smart Agriculture for Smallholder Farmers in Southern Africa. Faculty of Science and Agriculture, University of Fort Hare (2018)
17. Mccarthy, N., Branca, G., Lipper, L.: Climate smart agriculture: small holder adoption and implications for climate change adaptation and mitigation. Working Paper (2011). https://www.researchgate.net/publication/265229129
18. Mckinley, J.: Climate-Smart Agriculture (CSA) Technologies in Asia CCAFS Workshop Report Climate-Smart Agriculture Technologies in Asia Regional Workshop Report Workshop Organizers United Nations Environment Programme (UNEP) CGIAR Research Program on Climate Change, Agriculture, and Food Security (CCAFS) International Rice Research Institute (IRRI) Alabang and LosBaños, Philippines (2015)

19. Middelberg, S.L.: Sustainable agriculture: are views of challenges facing the South African agricultural sector. J. Hum. Ecol. **42**(2), 163–169 (2013)
20. Mungai, C., Nyasimi, M., Kimeli, P., Sayula, G., Radeny, M., Kinyangi, J.: Climate-smart agriculture technologies and practices for climate-resilient livelihoods in Lushoto, Northeast Tanzania. Climate **5**, 1–2 (2017). https://doi.org/10.3390/cli5030063
21. Mutoko, M.C.: Adoption of climate-smart agricultural practices: barriers, incentives, benefits and lessons learnt from the MICCA Pilot Site in Kenya. MICCA Programme, Food and Agriculture Organization (2014)
22. Mwongera, C., Shikuku, K.M., Twyman, J., Läderach, P., Ampaire, E., Van Asten, P., Winowiecki, L.A.: Climate smart agriculture rapid appraisal (CSA-RA): a tool for prioritizing context- specific climate smart agriculture technologies. Agric. Syst. **151**, 192–203 (2017). https://doi.org/10.1016/j.agsy.2016.05.009
23. Neufeldt, H., Jahn, M., Campbell, B.M., Beddington, J.R., Declerck, F., Pinto, A.D., Lezaks, D.: Beyond climate-smart agriculture: toward safe operating spaces for global food systems. Agric. Food Secur. **12**(2), 1–6 (2013)
24. Parry, M.: Climate change is a development issue, and only sustainable development can confront the challenge. Climate Dev. **1**(1), 5–9 (2009)
25. Phiri, M.A.R., Chilonda, P., Manyamba, C.: Challenges and opportunities for raising agricultural productivity in Malawi. Int. J. Agric. For. **2**(5), 210–224 (2012)
26. Prasannath, V.: Trends and developments in Sri Lanka's livestock industry. J. Stud. Manage. Planning **1**(4), 46–55 (2015)
27. Qui, C., Period, H., Wageningen, P.B.: Smallholder Farmer Perceptions on the Climate Smart Agriculture (CSA) Practices in Eastern India, Bihar. Farming System Ecology Group, Netharlands (2016)
28. Sain, G., María, A., Corner-dolloff, C., Lizarazo, M., Nowak, A., Martínez-barón, D., Andrieu, N.: Costs and benefits of climate-smart agriculture: the case of the Dry Corridor in Guatemala. Agric. Syst. **151**, 163–173 (2017)
29. Shija, S.M.Y.: Smallholder farmers' practices and understanding of climate change and climate smart agriculture in the Southern Highlands of Tanzania. J. Resour. Dev. Manage. **13**, 37–47 (2015)
30. Steenwerth, K.L., Hodson, A.K., Bloom, A.J., Carter, M.R., Cattaneo, A., Chartres, C.J., Jackson, L.E.: Climate-smart agriculture global research agenda: scientific basis for action. Agric. Food Secur. **11**(3), 1–39 (2014)
31. UNDP: Fighting climate change: human solidarity in a divided world. Human Development Report, 2007/08, Macmillan, New York, USA (2007)
32. Wickremasinghe, S.I.: Development of the National Agricultural Research System (NARS) in Sri Lanka with special reference to food crops sub-sector: issues related to science policy. J. Nat. Sci. Foundation Sri Lanka **34**(2), 69–83 (2006)
33. World Bank; CIAT: Climate-smart agriculture in Sri Lanka. In: CSA Country Profiles for Africa, Asia, and Latin America and the Caribbean Series. The World Bank Group, Washington D.C. (2015)

Chapter 10
Image Processing: A Smart Technology for Early Detection of Crop Pests and Diseases

Kandiah Pakeerathan

Introduction

Agriculture is the backbone of any developing country's economy and a major contributor to its GDP. There is an 80% probability that the world population will hit 9.6 billion in the year 2030, and will constantly demand to increase the current food production by 2% every year [26]. The arable land for crop cultivation is limited; therefore, the only way to overcome hunger is to increase the productivity per unit land area using high-yielding varieties. Invasion, evolution, and from time to time the emergence of catastrophic pest and disease outbreaks are constantly posing threat to the agriculture sector. According to FAO statistics, 20–40% of the global crop yield is being lost due to the damage caused by the pests and diseases. In India alone, 18% of the total crop production is being lost every year, and an estimated monetary value of Rs. 60,000 Crores. If all forms of crop loss are avoided, it would be enough to feed additional 100 million people annually.

These undesirable huge crop losses can be avoided if crop pests and diseases are detected in advance. Farmer's knowledge on accurate detection of the pest and diseases is minimal especially those living in developing and least developing countries, and without proper diagnosis and plant protection expert's guidance, farmers indiscriminately apply dangerous pesticides which can cause deleterious impact on human health and biodiversity if applied frequently [11].

K. Pakeerathan (✉)
Department of Agricultural Biology, Faculty of Agriculture, University of Jaffna, Jaffna, Sri Lanka
e-mail: pakeerathank@univ.jfn.ac.lk

© Centre for Science and Technology of the Non-aligned and Other Developing Countries 2023
K. Pakeerathan (ed.), *Smart Agriculture for Developing Nations*, Advanced Technologies and Societal Change, https://doi.org/10.1007/978-981-19-8738-0_10

135

Traditional Means of Pest and Disease Detection

Integrated pest management emphasizes the accurate and timely identification and control of weeds, insects, and diseases. Increasing pest's population and diseases causing pathogen inoculum potential above threshold levels may lead to a decline in the general health of agricultural and horticultural crops. If such problems are left untreated for too long, they may cause substantial economic damage and loss which will end up with food scarcity and insecurity [14]. Traditional way of pest identification and disease diagnosis can be difficult and often require consultation with a plant protection specialist. If the farm is far distant from the expert area, suspected samples need to be collected without disturbance, and deliver them to a specialist for identification [76]. Regular package delivery methods can take days, leading to delays in pest control recommendations. Sometimes, mailed samples deteriorate during shipment and become unsuitable for diagnosis. Farmers are expecting fast and accurate pest and disease detection methods to escape from disasters.

Thereafter, advances in Information Technology (IT) developed the Distance Diagnostic and Identification System (DDIS) for quick identification of pests and diseases. DDIS allows users to submit digital images obtained in the field for rapid diagnosis and identification of pest insects, weeds, diseases, and animals. Therefore, DDIS process had a fairly speed, reduced the samples collection and delivery time. The process of DDIS has four steps:

1. Observe a pest, unknown plant, or symptom,
2. Capture images through a digital camera with or without the use of a microscope,
3. Submit the sample, with the option to revise the sample after initial submission and
4. Await diagnosis by a clinic or specialist [76].

This system provides a digital image library with associated GPS location, crop, and pest or disorder data that can be used in future educational programs.

This method was also dependent on experts for the proper identification of pests and diseases from the captured picture. Farmers and commercial growers preferred to have a highly automated method that should not rely on the experts to detect the pest and diseases. Automated technology is not only important to take needful remedies on the spot, but also to develop a pest and disease forecasting and early warning system for prevention before the attack [68]. For example, the locust watch and migration forecasting system safeguards millions of tons of food and feed in the Indian subcontinent and African countries. Accurate and rapid diagnosis further can help producers to avoid costly mistakes (treating for the wrong pest, applying a fungicide when no fungicide is warranted, etc.).

Image Processing an Emerging Technology for Pest and Disease Detection

Now the world is in the cutting-edge technology era. The agriculture modernization projects are running all around the developed and developing nations to incorporate feasible and applicable smart agriculture technologies such as Global Positioning System (GPS) navigation of tractors, robotics, remote sensing, data analytics, unmanned aerial and terrestrial vehicles, drones, automated irrigation systems, biosensors, nanosensors, real-time monitoring, and prediction systems practically to find effective solution to existing problems of pest and diseases, water scarcity, drought, etc., [35, 72]. Among the identified technologies, image processing technique is currently being used intensively in pest and disease monitoring, detection and diagnosis, and management [44], because of the reliability improvement in the fields of computer vision and image processing, availability of the precision and accuracy of image analysis softwares [22]. The principle behind the image analysis technique is to distinguish the objects from the plant-based background, thereby isolated accurate feature information is used to make decisions by anybody in-ground scenario just using a smartphone with a quality camera and internet.

The standard steps of the image processing are.

1. Clear image capture
2. Resize the images
3. Color conversion
4. Segmentation
5. Reduction of noise
6. Pest and disease detection
7. Feature extraction
8. Feature selection
9. Image classification
10. Evaluation of pest and disease categories, and
11. Performance measures.

Clear image capture is the foremost important step in any image processing technique. In the evolutionary path of computer vision and machine learning, RGB, fluorescent, thermal, hyperspectral, and 3D images are used according to the reliability, feasibility, and cost [77]. There are various types of cameras are being used to capture sharp and high-resolution images in different field scenarios [7].

1. RGB/CIR cameras: Capture a combination of color RGB (red, green, and blue light) (called) or visible and infrared (CIR) imagery that enables the estimation of green biomass (NDVI type of information).
2. Hyperspectral (Microhyperspec VNIR) camera: The hyperspectral visible and near-infrared imager enables the acquisition of hundreds of images within the entire electromagnetic spectrum between the visible and the near-infrared wavelengths ranging from 400 to 900 nm in a continuous mode.

3. Long-wave infrared cameras or thermal imaging cameras: Thermal images obtained for phenotyping from aerial platforms [unmanned aerial vehicles (UAVs)] using thermal MIDAS cameras have a resolution in the range of 20–40 cm. Potential use of thermal imaging includes predicting water stress in crops, disease, and pathogen detection in plants, evaluating the maturing of fruits, and bruise detection in fruits and vegetables.
4. Multispectral cameras: Widely used to capture multispectral imagers with a limited number of spectral bands that are used for crop monitoring via remote sensing.
5. Conventional digital cameras (NDVI type work): Conventional digital cameras are low-cost instruments that enable to capture of live images of insect pests and disease symptoms.

The automated system containing software will do all the processes once the clear image loads and will give the output within a minute. In the early stages of image processing, classical and hand-crafted feature extraction methods were intensively used to pre-process the captured images (image resizing, filtering, color space conversion, and histogram equalization) [22, 25]. Then feature extracted images were trained using shallow classifier algorithms such as support vector machines (SVM), Naive Bayes (NB), principle component analysis (PCA), maximum likelihood classification (MLC), K-Nearest-Neighbors (KNN), decision trees (DT), random forest (RF), and artificial neural networks (ANN) [29, 43, 44, 53]. Then whichever the algorithm used for image training, were used to recognize features from new images captured from the field. The accuracy of this method varied from 75 to 90%. Thereafter, researches were focused on deep learning convolutional neural network (CNN) architectures to automatically perform feature extraction and image classification to using very sophisticated ICT tools, for example, multicore Graphics Processing Units (GPUs) and to train the large datasets and to increase the accuracy of the results more than 95% [44, 57]. In the deep learning CNN, VGG-16, VGG-19, ResNet-50, ResNet50V2, ResNet101V2, Inception-V3, Xception, MobileNet, SqueezeNet, AlexNet, VGG-Net and GoogleNet, etc., models are being used to pre-train the images to increase the image classification performances [11].

Using the automated image processing pipeline shown in Fig. 10.1, several insect pests and diseases are being precisely detected at their early infestation and infection stages, respectively, and with the immediate recommendation given by the software at various accuracy levels (Table 10.1). To increase the accuracy of image detection, recently developed deep transfer CNN models are highly being applied in the various ground scenarios as they are able to transfer the learned knowledge from one domain to another with the fewer number of images [65, 78]. The detected information will be helpful to develop an early warning and pest forecasting system to advise the farmers to protect their crops from pests and diseases and to implement exact control strategies [30, 70].

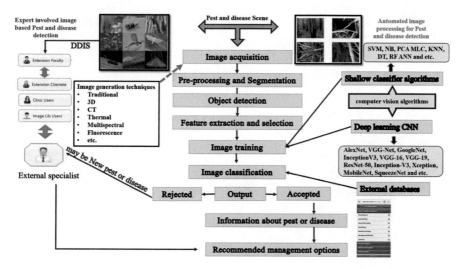

Fig. 10.1 Success model for pest and disease detection through image processing

Application of Image Processing in Smart Agriculture

Visual observation and qualitative and quantitative estimation of the damage caused by the pests and diseases are called 'Phenotyping'. Phenotyping is the first step in quantifying disease or pest incidence prior to unraveling the genetic control of traits of ecological and economical significance. Traditional phenotyping was laborious, costly, and time-consuming in large fields. Recent advances in computer vision and automation technology opened an image-based high-throughput phenotyping platform (HTP) that can process the images of at least hundreds to thousands of plants daily [61]. The fast development of improved varieties that are resistant to pests and diseases, and early detection and management of pests and diseases remarkably rely on monitoring, high-resolution, and high-throughput field-scale phenotyping technologies that can efficiently discriminate affected or damaged plants within a larger population across multiple environments [17, 77]. Very recent advances in HTP platforms coupled with Artificial intelligence (AI) techniques are monitoring very precisely single leaves/plant organs, individual plants, field plots, and full fields as required and record ultra-high-resolution data (> 1000 points/m^2) for phenotyping and studying biological processes at the individual tree level in ways never before imagined and processing the data very quickly to take the immediate solution from proximal to remote sensing. These varieties of platforms include environmental sensor networks, autonomous ground vehicles/rovers (www.terra-boost. com), phenomobiles/tractors/buggies, phenotyping towers, field scanning platforms (terraref.org), Drone phenotyping, and machine learning are coupled with artificial intelligence (UAVs), aircraft, zeppelins, satellites, etc. [17, 46, 59].

Image-guided pesticide spray is an electrostatic spray-based technology to precisely and effectively apply pesticide only to the target. In this technique, the

Table 10.1 Summary of pests and diseases detected using different deep and shallow image classification models/features

Pest/disease/symptom	Model/features	Accuracy	References
Vegetable Pests: Armyworm, Yellow stem borer, Snake gourd Fruit fly, *Riptortuspedestris*, Spotted beetle, Brinjal Mealybug, Green horned caterpillar, Aulacophora beetle, Ash weevil, and Brinjal flea beetle	VGG16 and InceptionV3	97.5% and 99.8%	Vijayakanthan et al. [71] (accepted)
Paddy leaf diseases: leaf blight, brown spot, and leaf blast	Support vector machine	89.19%, 82.86%, and 89.19%	Bandara and Mayurathan [12]
Rice pest and diseases in leaves	ResNet50 and ResNet101V2	75.0 and 86.79	Burhan et al. [16]
Paddy leaf diseases	VGGNet	91.83% to 92.00%	Chen et al. [18]
Plant Village dataset (PVD):Apple Rust, Apple Scab, Bell Pepper Leaf Spot, Corn Blight, Corn Gray Leaf Spot, Corn Rust, Grape Black Rot, Potato Early Blight, Potato Late Blight, Squash Powdery Mildew, Tomato Bacterial Spot, Tomato Early Blight, Tomato Late Blight, Tomato Mold, Tomato Mosaic Virus, Tomato Septoria Leaf Spot Tomato Yellow Virus	GoogleNet and AlexNet	99.35%	Singh et al. [62]
Apple (scab, black rot, cedar apple rust), cherry (powdery mildew, cospora leaf spot) diseases	VGG16 model with ImageNet	98.20%	Lee et al. [33]
Bacterial Spot, Early Blight, Late Blight, Leaf Mold, Septoria Leaf Spot, Spider Mite, Target Spot, Tomato Mosaic Virus, and Yellow Leaf Curl Virus disease in tomato	S-CNN and F-CNN	92% to 98%	Sharma et al. [60]

(continued)

amount of pesticide wasted can be reduced significantly by only applying the pesticide to the plants in a targeted manner. Sensor cameras are detecting damage or insect or pest and apply only after clear recognition of the image. This can lead to cost-effective pest control in terms of materials and labor while protecting natural enemies and the environment.

Table 10.1 (continued)

Pest/disease/symptom	Model/features	Accuracy	References
12 classes of pests: Acrididae, *Anticarsia gemmatalis*, Coccinellidae, *Diabrotica speciosa*, *Edessa meditabunda*, *Euschistus heros* adult, *E. heros* nymph, Gastropoda, *Lagria villosa*, *Nezara viridula* adult, *N. viridula* nymph and *Spodoptera* spp.	Deep learning with UAV images	93.82%	Tetila et al. [66]
Bactrocera litifrons, *Bemisia tabaci*, *Chrysodeixis chalcites*, *Epilachna vigintiopunctata*, *Spodoptera litura*, *Helicoverpa armigera*, *Icerya aegyptiaca*, *Liriomyza trifolii*, *Tuta absoluta* and *Nesidiocoris tenuis*	DenseNet169	88.83%	Pattnaik et al. [47]
Flea beetle, June beetle, lady bug, squash bug, and tarnish plant bug	Convolutional neural network	65.17% to 83.08%	Suthakaran and Premaratne [64]
Banana leafspots (Black sigatoka, Yellow sigatoka), Xanthomonas wilt, Fusarium wilt and Corm weevil damage	ResNet50 and InceptionV2	90%	Selvaraj et al. [57]
Coryneum beijerinckii, Apricot monilialaxa, Walnut leaf mite, Peach sphaerolecanium prunastri, *Xanthomonas arboricola*, Peach monilialaxa, *Erwinia amylovora*, and Cherry myzuscerasi	Support vector machine	95.5	Türkoğlu and Hanbay [69]
Weeds: sisymbriumsophia and Procumbent Speedwell	AlexNet	98.92%	Dawei et al. [19]
Passion fruit scab and woodiness diseases	Support vector machine	79%	Dharmasiri and Jayalal [21]
Tea insect pests in leaves	Support vector machines (SVM)	85.5%	Deng et al. [20]
82 insect pests in the field	Deep convolutional neural network (DCNN)	90%	Wang et al. [73]
Leaf spot diseases in maize	Artificial neural networks	98%	Mohan et al. [37]

(continued)

Table 10.1 (continued)

Pest/disease/symptom	Model/features	Accuracy	References
Caterpillar Tutaabsoluta, Thrips, Early blight, Late blight, and powdery mildew	Combinations of classifiers	87.80%	Es-saady et al. [24]
Brown spot disease, leaf blast disease, and bacterial blight disease in paddy leaf	AdaBoost algorithm	83.3% to 93.3%	Mohan et al. [37]
Leaf spot, rust, leaf spot and Cercospora leaf spot in Alfalfa leaf	Vector machine technique	80%	Qin et al. [52]
White flies, Aphids, and Cabbage moth	Support vector machine	95	Rajan et al. [53]
Corn leaf diseases (Southern Leaf Blight, Southern Rust, Gray Leaf Spot, Holcus Spot, and Stewart's Wilt)	Gabor filter and visual features	90.04%	Mousavi et al. [41]
Yellow Vein Mosaic Virus diseases in Okra leaf	Naives Bayes	87%	Mondal et al. [40]
Tea leaf virus and fungal diseases	Fuzzy	95.7%	Billah et al. [15]
Early blight, late blight, and septoria leaf spot in tomato	Fuzzy	95%	Muthukannan and Latha [42]
Caterpillar Tutaabsoluta, downy mildew, and powdery mildew in vegetable crops	Combinations of classifiers	Class of whites = 94% Class of yellow = 93%	El Massi et al. [23]
Powdery mildew and early blight in tomato	Vector machine technique	99.5%	Mokhtar et al. [38]
Early blight disease tomato leaf	Colors descriptor	100%	Molina et al. [39]
Bacterial blight in rice leaf	Artificial neural networks	100%	Orillo et al. [45]
Yellow vein mosaic virus disease in okra	PSO	93.3%	Zhang et al. [80]
Grey mildew and bacterial blight in cotton leafs	Artificial neural networks	94%	Revathi and Hemalatha [55]
Rice leaf bacterial diseases	Fractal dimension, and chaos theory	96%	Surendrababu et al. [63]
Sugarcane ring, rust, and yellow spots in sugarcane leaf	Vector machine technique	80%	Ratnasari et al. [54]
Paddy pests	Support vector machine	90	Venugoban and Ramanan [70]
Paddy leaf diseases	Fourier descriptor	83%	Asfarian et al. [9]

(continued)

Table 10.1 (continued)

Pest/disease/symptom	Model/features	Accuracy	References
Olive leaf spot	Color segment	86%[FCM] 66%[KMC]	Al-Tarawneh [5]
Mite diseases Cotton leaf	Diseases spot area	94.79%	Zhihua et al. [84]
Rust, anthracnose, white rot, fruit rust, ascochyta spot, and witches broom in Jujube leaf	Artificial neural networks	91%, 89%, 94%, 84%, 73% and 81% respectively	Zhang et al. [82]
Downy mildew, powdery mildew, and anthracnose in grape leafs	Artificial neural networks	100%	Sannakki et al. [56]
Downy mildew and Anthracnose in Watermelon leaf	Artificial neural networks	75% and 76.9%	Kutty et al. [32]
Fungus and leaf spots in maize leaf	Orthogonal locally discriminant projection	94%	Zhang and Zhang [83]
Downy mildew, powdery mildew, and anthracnose in cucumber leaf	Minimum path evaluation theory	96%	Pixia and Xiangdong [50]
Bacterial disease, sunburn, early scorch, late scorch and Fungal disease in rose, beans, lemon, and banana leaf	Vector machine technique	94%	Arivazhagan et al. [8]
Cercospora leaf spot in sugar beet	Vector machine technique	99.07%	Zhou et al. [85]
Greasy spot, melanose, and scab in citrus leaf	Discriminant analysis	98.75%	Bandi et al. [13]
Brown spot, bacterial blight, rice blast, and sheath rot in rice leaf	Rules based theory	91.19%	Phadikar et al. [48]
Blight, sheath blight, and southern blight in maize	Vector machine technique	98%	Lu et al. [34]
Lepidopteran pest	Locality constrained linear coding (LLC) and classification and regression tree (CART)	95%	Zhu and Zhang [86]
Grapes and wheat rust, downy, and powdery mildew in leaves	Color, shape, and texture feature	94.29%	Wang et al. [74]
Cotton leaf spot	Color	98.1%	Revathi and Hemalatha [55]

(continued)

Table 10.1 (continued)

Pest/disease/symptom	Model/features	Accuracy	References
Nutrition disease, namely, nitrogen, potassium, and magnesium in oil palm	Vector machine technique	95%	Asraf et al. [10]
Citrus canker	AdaBoost algorithm	88%	Zhang and Meng [81]
Early scorch, cottony mild, ashen mold, tiny whiteness and late scorch in various plants leaf and fruits	Artificial neural networks	94.67%	Al-Hiary et al. [4]
Brown spot in Cassava leaf	Artificial neural networks	79.23%	Abdullakasim et al. [2]
Anthracnose and frog eye diseases in tobacco	Artificial neural networks	88.593%	Guru et al. [27]
Early scorch, cottony mild, ashen mold, tiny whiteness and late scorch in various plants	Artificial neural networks	93%	Al Bashish et al. [6]
Cassava leaf mosaic disease	Neural networks and support vector machine	100%	Aduwo et al. [3]
Powdery mildew, leaf rust, leaf blight, and stripe rust in wheat	Combinations of classifiers	95.16%	Titan et al. [67]
Blast, brown narrow and brown spot disease in paddy crop leaf	Color and shape	94.7%	Kurniawati et al. [31]
Cucumber leafminer	Fractal dimension, and chaos theory	90%	Wu et al. [75]
Grape leaf scab and rust	Vector machine technique	97.8%	Meunkaewjinda et al. [36]
Cucumber leaf powdery mildew, and downy mildew	Vector machine technique	100%	Youwen et al. [79]
Leaf brown spot, blast, Sheath rot, and brown spot in rice	Artificial neural networks	94.21%	Phadikar and Sil [49]
Bacterial soft rot, bacterial brown spot, Phytophthora, and black rot diseases	Artificial neural networks	89.6%	Huang [28]
Corynespora, frog eye and Collectotrichum on rubber tree leaf	Artificial neural networks	96.5%	Abdullah et al. [1]
Citrus leaf diseases Melanose, greasy spot, and scab	Intensity texture features	96%	Pydipati et al. [51]
Fall armyworm in maize leaf	Contrasting lighting conditions	94.72%	Sena et al. [58]

Key factors that need to be considered while designing advanced image processing systems are as follows [30, 44].

1. Quality (resolution) of the pest or disease images.
2. Number of quality images used in machine learning.
3. A large number of dataset needs to be considered in the large amounts.
4. Need to acquire highly contrast images to prevent the background effect and noises.
5. Segmenting the exact spot in a plant part into meaningful disease.
6. Preparation of training and testing samples from the input image.
7. Entomologist or pathologist assistance for clear classification in recognizing segmented spots into meaningful images.
8. Take photos in all the environments and directions to prevent the color and contrast effect when the climate changed.
9. Select the photos of live, processed, and museum specimens to increase the accuracy.

Image processing software or application developers suffers from serious challenges including less contrast nature of symptoms and pests, the complexity of the wild environment, detection of tiny size pest, and classification of multiple classes of pests. While recent deep learning-based mobile vision techniques have shown some success in overcoming the above issues, one key problem is that toward large-scale multiple species of pest data, imbalanced classes significantly reduce their detection and recognition accuracy.

Conclusions and Future Prospects

Pests and diseases are major hurdles to agriculture. Fast and accurate detection and diagnosis are highly being preferred by commercial farmers to prevent the further advancement of pest and disease outbreaks. In the digital era, SI-coupled image processing system including mobile applications coupled databases, is a blooming technique for fast and rapid detection, diagnosis, and proposes suitable sustainable management options within a minute and prevent from over-application of unwanted nuisance pesticides which can cause health and environmental damages. Several image classification models from shallow to deep CNN have been used to detect the pest precisely, but every method has its own advantages and disadvantages. Even several cutting-edge techniques have been discovered in machine vision technology, still, a long way to go to overcome the obstacles like differentiate pests showing the similar color of plant surfaces, beneficial insects and pests showing similar morphology, and detection of the pathogen before expressing symptoms. Continuous efforts of scientists and new inventions in machine learning vision technology will overwhelm the obstacles encountered near future.

References

1. Abdullah, N.E., Rahim, A.A., Hashim, H., Kamal, M.M.: Classification of rubber tree leaf diseases using multilayer perceptron neural network. In: 2007 5th Student Conference on Research and Development, pp. 1–6. IEEE (2007)
2. Abdullakasim, W., Powbunthorn, K., Unartngam, J., Takigawa, T.: An images analysis technique for recognition of brown leaf spot disease in cassava. Tarım Makinaları Bilimi Dergisi **7**(2), 165–169 (2011)
3. Aduwo, J.R., Mwebaze, E., Quinn, J.A.: Automated vision-based diagnosis of cassava mosaic disease. In: Industrial Conference on Data Mining-Workshops, pp. 114–122. New York, NY (2010)
4. Al-Hiary, H., Bani-Ahmad, S., Reyalat, M., Braik, M., Alrahamneh, Z.: Fast and accurate detection and classification of plant diseases. Int. J. Comput. Appl. **17**(1), 31–38 (2011)
5. Al-Tarawneh, M.S.: An empirical investigation of olive leave spot disease using auto-cropping segmentation and fuzzy C-means classification. World Appl. Sci. J. **23**(9), 1207–1211 (2013)
6. Al Bashish, D., Braik, M., Bani-Ahmad, S.: A framework for detection and classification of plant leaf and stem diseases. In: 2010 International Conference on Signal and Image Processing, pp. 113–118. IEEE (2010)
7. Araus, J.L., Cairns, J.E.: Field high-throughput phenotyping: the new crop breeding frontier. Trends Plant Sci. **19**(1), 52–61 (2014). https://doi.org/10.1016/j.tplants.2013.09.008
8. Arivazhagan, S., Shebiah, R.N., Ananthi, S., Varthini, S.V.: Detection of unhealthy region of plant leaves and classification of plant leaf diseases using texture features. Agric. Eng. Int. CIGR J. **15**(1), 211–217 (2013)
9. Asfarian, A., Herdiyeni, Y., Rauf, A., Mutaqin, K.H.: Paddy diseases identification with texture analysis using fractal descriptors based on fourier spectrum. In: 2013 International Conference on Computer, Control, Informatics and Its Applications (IC3INA), pp. 77–81. IEEE (2013)
10. Asraf, H.M., Nooritawati, M., Rizam, M.S.: A comparative study in kernel-based support vector machine of oil palm leaves nutrient disease. Proc. Eng. **41**, 1353–1359 (2012)
11. Ayan, E., Erbay, H., Varçın, F.: Crop pest classification with a genetic algorithm-based weighted ensemble of deep convolutional neural networks. Comput. Electron. Agric. **179**, 105809 (2020). https://doi.org/10.1016/j.compag.2020.105809
12. Bandara, D., Mayurathan, B.: Detection and Classification of Rice Plant Diseases using Image Processing Techniques (2021)
13. Bandi, S.R., Varadharajan, A., Chinnasamy, A.: Performance evaluation of various statistical classifiers in detecting the diseased citrus leaves. Int. J. Eng. Sci. Technol. **5**(2), 298–307 (2013)
14. Barbedo, J.G.A.: Detecting and classifying pests in crops using proximal images and machine learning: a review. AI **1**(2), 312–328 (2020)
15. Billah, M., Miah, M.B.A., Hanifa, A., Amin, R.: Adaptive neuro fuzzy inference system based tea leaf disease recognition using color wavelet. Commun. Appl. Electron. **3**(5), 1–4 (2015)
16. Burhan, S.A., Minhas, S., Tariq, A., Hassan, M.N.: Comparative study of deep learning algorithms for disease and pest detection in rice crops. In: 2020 12th International Conference on Electronics, Computers and Artificial Intelligence (ECAI), pp. 1–5. IEEE (2020)
17. Chawade, A., van Ham, J., Blomquist, H., Bagge, O., Alexandersson, E., Ortiz, R.: High-Throughput field-phenotyping tools for plant breeding and precision agriculture. Agronomy **9**(5), 258 (2019)
18. Chen, J., Chen, J., Zhang, D., Sun, Y., Nanehkaran, Y.A.: Using deep transfer learning for image-based plant disease identification. Comput. Electron. Agric. **173**, 105393 (2020). https://doi.org/10.1016/j.compag.2020.105393
19. Dawei, W., Limiao, D., Jiangong, N., Jiyue, G., Hongfei, Z., Zhongzhi, H.: Recognition pest by image-based transfer learning. J. Sci. Food Agric. **99**(10), 4524–4531 (2019)
20. Deng, L., Wang, Y., Han, Z., Yu, R.: Research on insect pest image detection and recognition based on bio-inspired methods. Biosys. Eng. **169**, 139–148 (2018). https://doi.org/10.1016/j.biosystemseng.2018.02.008

21. Dharmasiri, S.B.D.H., Jayalal, S.: Passion fruit disease detection using image processing. In: 2019 International Research Conference on Smart Computing and Systems Engineering (SCSE), 28–28 March 2019, pp. 126–133 (2019). https://doi.org/10.23919/SCSE.2019.884 2799

22. Dhingra, G., Kumar, V., Joshi, H.D.: Study of digital image processing techniques for leaf disease detection and classification. Multimedia Tools Appl. **77**(15), 19951–20000 (2018). https://doi.org/10.1007/s11042-017-5445-8

23. El Massi, I., Saady, Y.E., El Yassa, M., Mammass, D., Benazoun, A.: Serial combination of two classifiers for automatic recognition of the damages and symptoms on plant leaves. In: 2015 Third World Conference on Complex Systems (WCCS), pp. 1–6. IEEE (2015)

24. Es-saady, Y., El Massi, I., El Yassa, M., Mammass, D., Benazoun, A.: Automatic recognition of plant leaves diseases based on serial combination of two SVM classifiers. In: 2016 International Conference on Electrical and Information Technologies (ICEIT), pp. 561–566. IEEE (2016)

25. Geetharamani, G., Arun Pandian, J.: Identification of plant leaf diseases using a nine-layer deep convolutional neural network. Comput. Electr. Eng. **76**, 323–338 (2019). https://doi.org/10.1016/j.compeleceng.2019.04.011

26. Gerland, P., Raftery, A.E., Sevčíková, H., Li, N., Gu, D., Spoorenberg, T., Alkema, L., Fosdick, B.K., Chunn, J., Lalic, N., Bay, G., Buettner, T., Heilig, G.K., Wilmoth, J.: World population stabilization unlikely this century. Science. 346(6206):234–7 (2014). https://doi.org/10.1126/science.1257469

27. Guru, D., Mallikarjuna, P., Manjunath, S.: Segmentation and classification of tobacco seedling diseases. In: Proceedings of the Fourth Annual ACM Bangalore Conference, pp. 1–5 (2011)

28. Huang, K.-Y.: Application of artificial neural network for detecting Phalaenopsis seedling diseases using color and texture features. Comput. Electron. Agric. **57**(1), 3–11 (2007)

29. Kasinathan, T., Uyyala, S.R.: Machine learning ensemble with image processing for pest identification and classification in field crops. Neural Comput. Appl. 1–14 (2021)

30. Kumar, S.S., Raghavendra, B.: Diseases detection of various plant leaf using image processing techniques: a review. In: 2019 5th International Conference on Advanced Computing & Communication Systems (ICACCS), pp. 313–316. IEEE (2019)

31. Kurniawati, N.N., Abdullah, S.N.H.S., Abdullah, S.: Investigation on image processing techniques for diagnosing paddy diseases. In: 2009 International Conference of Soft Computing and Pattern Recognition, pp. 272–277. IEEE (2009)

32. Kutty, S.B., Abdullah, N.E., Hashim, H., Kusim, A.S., Yaakub, T.N.T., Yunus, P.N.A.M., Abd Rahman, M.F.: Classification of watermelon leaf diseases using neural network analysis. In: 2013 IEEE Business Engineering and Industrial Applications Colloquium (BEIAC), pp. 459–464. IEEE (2013)

33. Lee, S.H., Goëau, H., Bonnet, P., Joly, A.: New perspectives on plant disease characterization based on deep learning. Comput. Electron. Agric. **170**, 105220 (2020). https://doi.org/10.1016/j.compag.2020.105220

34. Lu, C., Gao, S., Zhou, Z.: Maize disease recognition via fuzzy least square support vector machine. J. Inf. Comput. Sci. **8**(4), 316–320 (2013)

35. Mat, I., Kassim, M.R.M., Harun, A.N., Yusoff, I.M.: Smart agriculture using Internet of Things. In: 2018 IEEE Conference on Open Systems (ICOS), 21–22 Nov 2018, pp. 54–59 (2018). https://doi.org/10.1109/ICOS.2018.8632817

36. Meunkaewjinda, A., Kumsawat, P., Attakitmongcol, K., Srikaew, A.: Grape leaf disease detection from color imagery using hybrid intelligent system. In: 2008 5th International Conference on Electrical Engineering/Electronics, Computer, Telecommunications and Information Technology, pp 513–516. IEEE (2008)

37. Mohan, K.J., Balasubramanian, M., Palanivel, S.: Detection and recognition of diseases from paddy plant leaf images. Int. J. Comput. Appl. **144**(12) (2016)

38. Mokhtar, U., Ali, M.A., Hassenian, A.E., Hefny, H.: Tomato leaves diseases detection approach based on support vector machines. In: 2015 11th International Computer Engineering Conference (ICENCO), pp. 246–250. IEEE (2015)

39. Molina, J.F., Gil, R., Bojacá, C., Gómez, F., Franco, H.: Automatic detection of early blight infection on tomato crops using a color based classification strategy. In: 2014 xix Symposium on Image, Signal Processing and Artificial Vision, pp. 1–5. IEEE (2014)
40. Mondal, D., Chakraborty, A., Kole, D.K., Majumder, D.D.: Detection and classification technique of yellow vein mosaic virus disease in okra leaf images using leaf vein extraction and Naive Bayesian classifier. In: 2015 International Conference on Soft Computing Techniques and Implementations (ICSCTI), pp. 166–171. IEEE (2015)
41. Mousavi, S., Hanifeloo, Z., Sumari, P., Arshad, M.M.: Enhancing the diagnosis of corn pests using gabor wavelet features and svm classification (2016)
42. Muthukannan, K., Latha, P.: Fuzzy inference system based unhealthy region classification in plant leaf image. Int. J. Comput. Inf. Eng. **8**(11), 2103–2107 (2015)
43. Nam, N.T., Hung, P.D.: Pest detection on traps using deep convolutional neural networks. In: Proceedings of the 2018 International Conference on Control and Computer Vision, pp. 33–38 (2018)
44. Ngugi, L.C., Abelwahab, M., Abo-Zahhad, M.: Recent advances in image processing techniques for automated leaf pest and disease recognition—a review. Inf. Process. Agric. (2020). https://doi.org/10.1016/j.inpa.2020.04.004
45. Orillo, J.W., Cruz, J.D., Agapito, L., Satimbre, P.J., Valenzuela, I.: Identification of diseases in rice plant (*Oryza sativa*) using back propagation artificial neural network. In: 2014 International Conference on Humanoid, Nanotechnology, Information Technology, Communication and Control, Environment and Management (HNICEM), pp 1–6. IEEE (2014)
46. Patrick, A., Pelham, S., Culbreath, A., Holbrook, C.C., Godoy, I.J.D., Li, C.: High throughput phenotyping of tomato spot wilt disease in peanuts using unmanned aerial systems and multi-spectral imaging. IEEE Instrum. Meas. Mag. **20**(3), 4–12 (2017). https://doi.org/10.1109/MIM. 2017.7951684
47. Pattnaik, G., Shrivastava, V.K., Parvathi, K.: Transfer learning-based framework for classification of pest in tomato plants. Appl. Artif. Intell. **34**(13), 981–993 (2020). https://doi.org/10. 1080/08839514.2020.1792034
48. Phadikar, S., Sil, J., Das, A.K.: Rice diseases classification using feature selection and rule generation techniques. Comput. Electron. Agric. **90**, 76–85 (2013). https://doi.org/10.1016/j. compag.2012.11.001
49. Phadikar, S., Sil, J.: Rice disease identification using pattern recognition techniques. In: 2008 11th International Conference on Computer and Information Technology, pp. 420–423. IEEE (2008)
50. Pixia, D., Xiangdong, W.: Recognition of greenhouse cucumber disease based on image processing technology. Open J. Appl. Sci. **3**(01), 27–31 (2013)
51. Pydipati, R., Burks, T., Lee, W.: Identification of citrus disease using color texture features and discriminant analysis. Comput. Electron. Agric. **52**(1–2), 49–59 (2006)
52. Qin, F., Liu, D., Sun, B., Ruan, L., Ma, Z., Wang, H.: Identification of alfalfa leaf diseases using image recognition technology. PLoS ONE **11**(12), e0168274 (2016)
53. Rajan, P., Radhakrishnan, B., Suresh, L.P.: Detection and classification of pests from crop images using support vector machine. In: 2016 International Conference on Emerging Technological Trends (ICETT), 21–22 Oct 2016, pp. 1–6 (2016). https://doi.org/10.1109/ICETT. 2016.7873750
54. Ratnasari, E.K., Mentari, M., Dewi, R.K., Ginardi, R.H.: Sugarcane leaf disease detection and severity estimation based on segmented spots image. In: Proceedings of International Conference on Information, Communication Technology and System (ICTS) 2014, pp 93–98. IEEE (2014)
55. Revathi, P., Hemalatha, M.: Cotton leaf spot diseases detection utilizing feature selection with skew divergence method. Int. J. Sci. Eng. Technol. **3**(1), 22–30 (2014)
56. Sannakki, S.S., Rajpurohit, V.S., Nargund, V., Kulkarni, P.: Diagnosis and classification of grape leaf diseases using neural networks. In: 2013 Fourth International Conference on Computing, Communications and Networking Technologies (ICCCNT), pp 1–5. IEEE (2013)

57. Selvaraj, M.G., Vergara, A., Ruiz, H., Safari, N., Elayabalan, S., Ocimati, W., Blomme, G.: AI-powered banana diseases and pest detection. Plant Methods **15**(1), 92 (2019). https://doi.org/10.1186/s13007-019-0475-z
58. Sena, D., Jr., Pinto, F., Queiroz, D., Viana, P.: Fall armyworm damaged maize plant identification using digital images. Biosys. Eng. **85**(4), 449–454 (2003)
59. Shakoor, N., Lee, S., Mockler, T.C.: High throughput phenotyping to accelerate crop breeding and monitoring of diseases in the field. Curr. Opin. Plant Biol. **38**, 184–192 (2017). https://doi.org/10.1016/j.pbi.2017.05.006
60. Sharma, P., Berwal, Y.P.S., Ghai, W.: Performance analysis of deep learning CNN models for disease detection in plants using image segmentation. Inf. Proces. Agric. **7**(4), 566–574 (2020). https://doi.org/10.1016/j.inpa.2019.11.001
61. Singh, A., Ganapathysubramanian, B., Singh, A.K., Sarkar, S.: Machine learning for high-throughput stress phenotyping in plants. Trends Plant Sci. **21**(2), 110–124 (2016). https://doi.org/10.1016/j.tplants.2015.10.015
62. Singh, D., Jain, N., Jain, P., Kayal, P., Kumawat, S., Batra, N.: PlantDoc: a dataset for visual plant disease detection. In: Proceedings of the 7th ACM IKDD CoDS and 25th COMAD, pp. 249–253 (2020)
63. Surendrababu, V., Sumathi, C., Umapathy, E.: Detection of rice leaf diseases using chaos and fractal dimension in image processing. Int. J. Comput. Sci. Eng. **6**(1), 69 (2014)
64. Suthakaran, A., Premaratne, S.: Detection of the affected area and classification of pests using convolutional neural networks from the leaf images. Int. J. Comput. Sci. Eng. (IJCSE) (2020)
65. Tan, C., Sun, F., Kong, T., Zhang, W., Yang, C., Liu, C.A.: Survey on deep transfer learning. In: International Conference on Artificial Neural Networks, pp. 270–279. Springer (2018)
66. Tetila, E.C., Machado, B.B., Astolfi, G., Belete, NAd.S., Amorim, W.P., Roel, A.R., Pistori, H.: Detection and classification of soybean pests using deep learning with UAV images. Comput. Electron. Agric. **179**, 105836 (2020). https://doi.org/10.1016/j.compag.2020.105836
67. Titan, Y., Zhao, C., Lu, S., Guo, X.: SVM-based multiple classifier system for recognition of wheat leaf diseases. In: Proceedings of 2010 Conference on Dependable Computing (CDC'2010) (2010)
68. Tonnang, H.E.Z., Hervé, B.D.B., Biber-Freudenberger, L., Salifu, D., Subramanian, S., Ngowi, V.B., Guimapi, R.Y.A., Anani, B., Kakmeni, F.M.M., Affognon, H., Niassy, S., Landmann, T., Ndjomatchoua, F.T., Pedro, S.A., Johansson, T., Tanga, C.M., Nana, P., Fiaboe, K.M., Mohamed, S.F., Maniania, N.K., Nedorezov, L.V., Ekesi, S., Borgemeister, C.: Advances in crop insect modelling methods—towards a whole system approach. Ecol. Model. **354**, 88–103 (2017). https://doi.org/10.1016/j.ecolmodel.2017.03.015
69. Türkoğlu, M., Hanbay, D.: Plant disease and pest detection using deep learning-based features. Turk. J. Electr. Eng. Comput. Sci. **27**(3), 1636–1651 (2019)
70. Venugoban, K., Ramanan, A.: Image classification of paddy field insect pests using gradient-based features. Int. J. Mach. Learn. Comput. **4**(1), 1 (2014)
71. Vijayakanthan, G., Kokul, T., Pakeerathan, K., Pinidiyaarachchi, U.A.J.: Classification of vegetable plant pests using deep transfer learning. In: 10th IEEE International Conference on Information and Automation for Sustainability (ICIAfS), Negombo, Sri Lanka, 11–13 Aug 2021 (2021) [Accepted]
72. Walter, A., Finger, R., Huber, R., Buchmann, N.: Opinion: smart farming is key to developing sustainable agriculture. Proc. Natl. Acad. Sci. **114**(24), 6148–6150 (2017). https://doi.org/10.1073/pnas.1707462114
73. Wang, R., Zhang, J., Dong, W., Yu, J., Xie, C.J., Li, R., Chen, T., Chen, H.: A crop pests image classification algorithm based on deep convolutional neural network. Telkomnika **15**(3), 1239–1246 (2017)
74. Wang, H., Li, G., Ma, Z., Li, X.: Image recognition of plant diseases based on principal component analysis and neural networks. In: 2012 8th International Conference on Natural Computation, pp. 246–251. IEEE (2012)
75. Wu, D.-k., Xie, C.-y., Ma, C.-w.: The SVM classification leafminer-infected leaves based on fractal dimension. In: 2008 IEEE Conference on Cybernetics and Intelligent Systems, pp. 147–151. IEEE (2008)

76. Xin, J., Buss, L., Harmon, C., Vergot, III P., Frank, M., W.L.: Plant and Pest Diagnosis and Identification Through DDIS. U.S. Department of Agriculture. https://edis.ifas.ufl.edu/ae225. Accessed 02 Jan 2021 (2002)
77. Yang, W., Feng, H., Zhang, X., Zhang, J., Doonan, J.H., Batchelor, W.D., Xiong, L., Yan, J.: Crop phenomics and high-throughput phenotyping: past decades, current challenges, and future perspectives. Mol. Plant **13**(2), 187–214 (2020). https://doi.org/10.1016/j.molp.2020. 01.008
78. Yin, H., Gu, Y.H., Park, C.-J., Park, J.-H., Yoo, S.J.: Transfer learning-based search model for hot pepper diseases and pests. Agriculture **10**(10), 439 (2020)
79. Youwen, T., Tianlai, L., Yan, N.: The recognition of cucumber disease based on image processing and support vector machine. In: 2008 Congress on Image and Signal Processing, pp. 262–267. IEEE (2008)
80. Zhang, Z., Li, Y., Wang, F., He, X.: A particle swarm optimization algorithm for neural networks in recognition of maize leaf diseases. Sens. Transducers **166**(3), 181 (2014)
81. Zhang, M., Meng, Q.: Automatic citrus canker detection from leaf images captured in field. Pattern Recogn. Lett. **32**(15), 2036–2046 (2011)
82. Zhang, W., Teng, G., Wang, C.: Identification of jujube trees diseases using neural network. Optik **124**(11), 1034–1037 (2013)
83. Zhang, S., Zhang, C.: Orthogonal locally discriminant projection for classification of plant leaf diseases. In: 2013 Ninth International Conference on Computational Intelligence and Security, pp. 241–245. IEEE (2013)
84. Zhihua, D., Huan, W., Yinmao, S., Yunpeng, W.: Image segmentation method for cotton mite disease based on color features and area thresholding. J. Theor. Appl. Inf. Technol. **48**(1) (2013)
85. Zhou, R., Tanaka, F., Kayamori, M., Shimizu, M.: Matching-based Cercospora leaf spot detection in sugar beet. Int. J. Nutr. Food Eng. **7**(7), 712–718 (2013)
86. Zhu, L.-q., Zhang, Z.: Using CART and LLC for image recognition of Lepidoptera. The Pan-Pacific Entomol. **89**(3), 176–186, 111 (2013)

Chapter 11
Development of a Smart Sprayer for Smallholder Farmers in Conservation Agriculture

T. Yu, C. Pretorius, and J. van Biljon

Introduction

Conservation agriculture (CA) is broadly referred as any soil management with reduced tillage, minimized tillage and no till in terms of mechanization, together with permanent soil cover and crop diversification/rotations [4]. Over years, CA has been reported to benefit farming practices with sustainability, enhanced biodiversity, carbon sequestration, labor savings, healthier soils, and ultimately increased yields in many countries [5].

Dry-land and rain-fed maize grain production is the dominant farming practice in the majority of summer rainfall areas in South Africa. A typical agricultural system in such practice includes an annual single mono crop of maize or in rotation with soybean or sunflower in recent years as part of CA. No-till and minimum tillage are most popular conservation tillage approaches that have been widely adopted by South African pioneer CA farmers from around 20 years ago [3,18]. In comparison with conventional farming, conservation tillage has been promoted by bringing major advantages such as reduced energy consumption and labor cost, decreased soil disturbance and erosion, and improved soil moisture reservation.

However, many gains from CA will take much longer time to establish themselves after some challenges are dealt with properly such as weed control and soil compaction [16]. According to many farmers who are practicing no till and researchers who are studying the impact of CA, the weed control is always one

T. Yu (✉) · C. Pretorius · J. van Biljon
ARC-Natural Resources and Agricultural Engineering, Pretoria, South Africa
e-mail: YuT@arc.agric.za

C. Pretorius
e-mail: PretoriusC@arc.agric.za

J. van Biljon
e-mail: vBiljonJ@arc.agric.za

© Centre for Science and Technology of the Non-aligned and Other Developing Countries 2023
K. Pakeerathan (ed.), *Smart Agriculture for Developing Nations*, Advanced Technologies and Societal Change, https://doi.org/10.1007/978-981-19-8738-0_11

151

of the top challenges and consequently crucial to sustainability of the CA farming practice in the early years after implementation of CA [9, 20].

Furthermore, weed management cost by applying chemical herbicide is also crucial for gaining adequate yields to compensate for added total cost in the beginning years after CA implementation and thereafter [11, 12].

Large-scale commercial CA systems in South America and North America where the CA has been pioneered, weed management through a combination of several herbicides, representing a significant portion of the variable costs of agricultural production [2]. In African context, one of the main challenges of herbicide application for smallholder farmers in CA is the lack of access to basic inputs and cash by smallholder farmers [1, 6]. Women and female-led households in particular are even more disadvantaged when accessing herbicides due to their status and role in fund allocation in the households, thereby reducing their ability to effectively use new technologies [13]. Thus, CA adoption in southern Africa may be dependent on affordability of herbicides for smallholder farmers and any effort on reducing the application of herbicide is greatly helpful for the smallholder farmers on saving cost and improving household fiscal [17].

All the herbicides currently available are categorized as pre-emergence or post-emergence, but most are applied directly to all soil surfaces no matter if the weeds are present or not. The adverse environmental impacts of accelerated herbicide use, particularly the glyphosate-based (Roundup), have been well reported since 20 years ago [8, 14, 19]. It is popularly concerned that while herbicide use has contributed vastly in weed control under CA farming, irresponsible and overuse of herbicides has been increasing the risk of contaminating the soil and ground/underground water in a long term and is a threat to environment and natural resources. Therefore, herbicide application must be done cautiously and reduced to minimal [15].

In order to limit overapplication on herbicides for weed control under CA systems to reduce cost and curb environmental adverse, numerous alternative methods have been investigated. Among the investigated solutions, one of the most promising is the technology-driven precision and smart spraying systems which have been gaining great attention by researchers and manufacturers [10]. In the past few years, some precision and smart spraying systems based on computer vision and imagery analysis have been already available at commercial farming level in the market, but the price is still beyond the affordability of the commercial farmers in many countries. Nevertheless, development of a precision and smart sprayer based on appropriate technologies for small-scale farmers is becoming urgent to expedite the transformation process from conventional to CA farming.

Therefore, the aim of the research was set to develop a cost-effective, small-scale, intelligent spraying system for smallholder farmers in CA. The target groups are the smallholder farmers with technical/expert assistance of the researchers. The aim of the project is to be achieved through design, laboratory testing and field evaluation of a low cost small-scale smart sprayer prototype for CA based on automation and precision concept.

The specific objectives of the project were:

1. To develop a cost-effective, small-scale, intelligent spraying system for smallholder farming in CA to apply on-target spot spraying of herbicide for weed control; and to design, laboratory test, and in-field evaluate a low-cost small-scale spot sprayer prototype.
2. The goal of the sprayer development is mainly to save costs for the farmers when weed control is needs under the circumstances that spots of weeds are appearing. Spot spraying is expected to save chemicals such as herbicides up to 80% when spraying only applied where weeds are present.
3. There is an expectation that the developed and evaluated prototype sprayer will be used to apply similar chemicals such as fungicides, insecticides, and other products for biological control.

Materials and Methods

The research undertaken followed a typical prototype design and evaluation process.

Concept Design

The concept of the prototype was that based on moderate technologies and previous technical work, the cost-effective small-scale smart spraying system to be designed would spray proper amount of chemical mixture, i.e., the herbicide mixture onto the targeted vegetation only in order to eliminate the weeds and avoid the majority of the soil surface to be sprayed. To achieve this, the weed-detecting sensor was used to see the weeds, and the spraying time and amount were determined by a programmable microchip controller and precision nozzle according to the spot size of the weeds spreading. The prototype was expected to be suitable for weed control at early stage of post-emergency weeds only.

The designed concept of the prototype spraying system is explained in the flowchart as shown in Fig. 11.1.

The designed concept prototype spraying system mounted on a tractor under operational conditions is shown in Fig. 11.2.

Function Design and Major Components

To make the spraying system programmable and controllable in timing and quantity, the parameters must be in or possibly converted into digital format variables. Therefore, the components and the system were selected as DC-powered for convenience of control and easy maintenance. Due to the fact that the auto-controlled sprayer

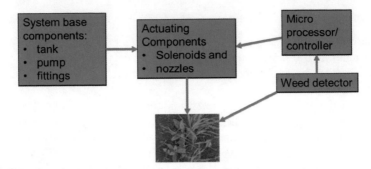

Fig. 11.1 The flowchart explaining the designed concept of the prototype spraying system

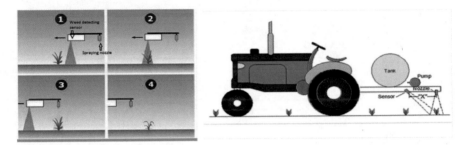

Fig. 11.2 The designed concept prototype spraying system mounted on a tractor under operational conditions

prototype was designed and modified from a previous manually controlled model on which the pump and pressure regulator were already tested [7], the same relevant components were used.

To make the system automatic spraying, a proper weed-detecting sensor was identified as critical component. Various types of weed-detecting sensors including laser, ultrasonic, and optical sensors have been tested for the suitability of this specific purpose at this stage.

The major components and functions are shown in Fig. 11.3 and Table 11.1.

The solenoids with precision spraying nozzles were powered 12 V DC and sponsored by a local distributor.

The controller of the prototype was developed based on microchip technique so that as soon as the weeds were seen by sensor, the solenoid was instructed by the controller to control opening and closing of the nozzle to keep spraying during the period when the spreading spot of weed was in detection.

Practically the detecting sensor and the solenoid with nozzle ought to be mounted horizontally at different locations, so that the delayed response time from the solenoid to spraying could be compensated. The suitable horizontal distance between the mounting positions needed to be determined by solenoid response time and the

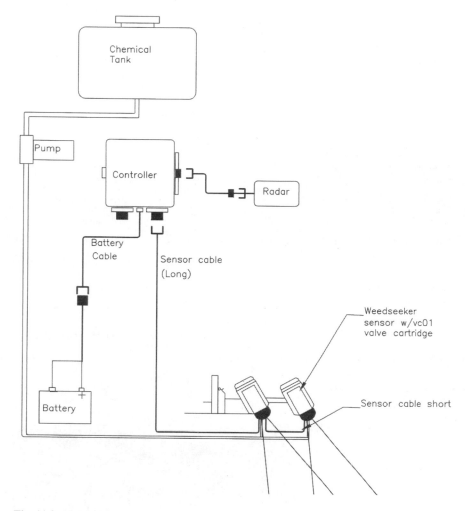

Fig. 11.3 Illustration of spraying system and major components

tractor ground speed. The calculated time delay between sensor and spraying nozzle is shown in Table 11.2.

The components and the assembled system were pre-programmed and tested on desktop (Fig. 11.4) before the field evaluation.

Table 11.1 List of major components and functions

Item	Quantity	Specification	Function
Battery	1	DC 12 V	Power source
Pump	1	DC 12 V	Pumping
Pressure regulator	1	DC 12 V	Stabilizing of pressure
Storage tank	1	Plastic with filters	Chemical storage
Radar	1	Ground speed radar with GPS	Speed sensor and control of flow rate
Weed detecting sensor	1/per spraying unit	12 V	Detecting of weed and output signal to controller
Controller	1	Arduino Uno	Controlling and processing
Solenoid and spray nozzle	1/per spraying unit	TeeJet XR range	Control of spray

Table 11.2 Determined distance between the sensor and nozzle due to spraying delay time and travel speed

Spraying delay time between the sensor and nozzle			
Travel speed (km/h)	Travel speed (m/s)	Sensor to nozzle distance (m)	Spraying delay time (s)
6	1.67	0.5	0.30
7	1.94	0.5	0.26
8	2.22	0.5	0.23

Fig. 11.4 Pre-programming and testing of components

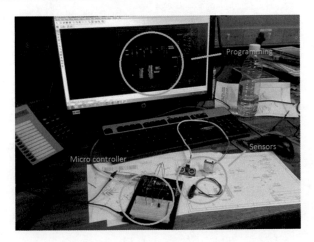

Prototype Construction

Based on the concept design, a single spray unit as the first prototype was constructed by utilizing and modifying an existing boom sprayer as shown in Fig. 11.5. The components mounted on the sprayer are as shown in Fig. 11.6.

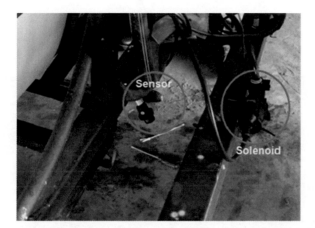

Fig. 11.5 A single unit of the prototype sprayer with sensor, solenoid, and nozzle installed by using a boom sprayer

Fig. 11.6 Practical tests conducted on solid ground

Selection of Weed-Detecting Sensors

It was expected that the weed detection was the bottleneck of the project and three low-cost sensors were selected to test with the prototype sprayer, namely, the ultrasonic object detector, laser object detector, and color detector.

Practical Tests

A series of tests were conducted after the functions of the spraying system were verified after assembling. The prototype with different detecting sensors was tested practically on solid ground with real weeds collected from field to evaluate if the sensor detects or distinguishes the weed from ground as shown in Fig. 11.6.

Results and Discussion

From the practical tests repeated 10 times by each setup, the findings regarding the effectiveness of the sensors are listed in Tables 11.3 and 11.4. (Two vertical ground distances were tested, i.e. 30 cm and 50 cm, respectively).

The following results can be summarized as following from the evaluation tests:

- The sensors had better effectiveness in shortened vertical ground distance, i.e., 30 cm better than 50 cm.

Table 11.3 Detecting effectiveness of the sensors (detector mounted 50 cm above ground level)

Detector type	Detecting effectiveness according to dimensions of weed spreading	
	Tall and thin (20 cm × 10 cm)	Low and wide (5 cm × 30 cm)
Laser	70%	30%
Ultrasonic	10%	40%
Color detector	0%	10%

Table 11.4 Detecting effectiveness of the sensors (detector mounted 30 cm above ground level)

Detector type	Detecting effectiveness according to dimensions of weed spreading	
	Tall and thin (20 cm × 10 cm)	Low and wide (5 cm × 30 cm)
Laser	80%	30%
Ultra sonic	20%	50%
Color detector	20%	20%

- The color sensor had the lowest detection rate in all setups.
- The best result was from the laser detector with 80% rate at 30 cm.
- The solenoid responds effectively.
- The responding time needed regular calibration before operation.

In this study, two categories of sensors have been tried, i.e., imagery sensor (color sensor) and range-sensors (laser and ultrasonic). The recommended identification rate of spray area for tree fruit is above 92%, the detecting effectiveness as achieved from this study is rather unsatisfactory [10].

Conclusions

A prototype low-cost smart sprayer was developed with low-cost and appropriate technologies through alternative modifications from previous research work. The major components were functioning as designed except the weed-detecting sensors. Therefore, for a smart spraying at the real-time recognition of weeds is crucially important. The developed prototype sprayer has potential to be used to apply similar chemicals such as fungicides, insecticides, and others. It is indicated that the identification of weeds is still at acceptably low efficiencies; therefore, the suggestion is that more future focused work will be on finding alternative and more efficient weed detectors. It is also recommended that improvement and evaluation will have to be done in the future especially on the fine tuning of the responding time and continued effort to improve the durability of the spraying system in field conditions.

References

1. Andersson, J.A., D'Souza, S.: From adoption claims to under-standing farmers and contexts: a literature review of conservation agriculture (CA) adoption among smallholder farmers in Southern Africa. Agric. Ecosyst. Environ. **187**, 116–132 (2014)
2. Bajwa, A.A.: Sustainable weed management in conservation agriculture. Crop Prot. **65**, 105–113 (2014)
3. Du Toit, G.: Promoting conservation agriculture in South Africa: a case study among commercial grain producers in the North West Province. The Bureau for Food and Agricultural Policy (BFAP) Report # 2007—04 (2007)
4. FAO: The status of conservation agriculture in Southern Africa: challenges and opportunities for expansion. REOSA Technical Brief 03, Food and Agriculture Organization of the United Nations Regional Emergency Office for Southern Africa (REOSA) (2010)
5. FAO: Exchanging Experience with Conservation Agriculture Technical guidelines. Food and Agriculture Organization of the United Nations (2014)
6. Giller, K.E., Witter, E., Corbeels, M., Tittonell, P.: Conservation agriculture and smallholder farming in Africa: the heretics' view. Field Crops Res. **114**(1), 23–34 (2009)
7. Hatting, J., Pretorius, C.: Report on the development of apesticide applicator for Rooibos, *Aspalathus linearis*. ARC Technical Report #PR 2012/0010 (2012)

8. Kumar, V., Singh, S., Chhokar, R.S., Malik, R.K., Brainard, D.C., Ladha, J.K.: Weed management strategies to reduce herbicide use in zero-till rice-wheat cropping systems of the Indo-Gangetic plains. Weed Technol. **27**(1), 241–254 (2013)

9. Lafond, G.P., McConkey, B.G., Stumborg, M.: Conservation tillage models for small-scale farming: linking the Canadian experience to the small farms of Inner Mongolia autonomous region in China. Soil Till. Res. **104**, 150–155 (2009)

10. Mahmud, M.S., Zahid, A., He, L., Martin, P.: Opportunities and possibilities of developing an advanced precision spraying system for tree fruits. Sensors **2021**(21), 3262 (2021)

11. Mavunganidze, Z., Madakadze, I.C., Nyamangara, J., Mafongoya, P.: The impact of tillage system and herbicides on weed density, diversity and yield of cotton (*Gossipium hirsutum* L.) and maize (*Zeamays* L.) under the smallholder sector. Crop Prot. **58**, 25–32 (2014)

12. Muoni, T., Rusinamhodzi, L., Thierfelder, C.: Weed control in conservation agriculture systems of Zimbabwe: identifying economical best strategies. Crop Prot. **53**, 23–28 (2013)

13. Nyanga, P.H., Johnsen, F.H., Kalinda, T.H.: Gendered impacts of conservation agriculture and paradox of herbicide use among small-holder farmers. Int. J. Technol. Dev. Stud. **3**(1), 1–24 (2012)

14. Ramdas, K., Gairhe, B., Kadyampakeni, D., Batuman, O., Alferez, F.: Glyphosate: its environmental persistence and impact on crop health and nutrition. Plants **8**, 499 (2019)

15. Rahman, M.M.: Potential environmental impacts of herbicides used in agriculture. J. Agric. Forest Meteorol. Res. **3**(1), 266–269 (2020)

16. Rusinamhodzi, L., Corbeels, M., Van Wijk, M.T., Rufino, M.C., Nyamangara, J., Giller, K.E.: A meta-analysis of long-term effects of conservation agriculture on maize grain yield under rain-fed conditions. Agron. Sustain. Dev. **31**(4), 657–673 (2011)

17. Smith, H.J., Kruger, E., Knot, J., Blignaut, J.: Conservation agriculture in South Africa: lessons from case studies. In: Kassam, A., Mkomwa, S., Friedrich, T. (eds.), Conservation Agriculture for Africa: Building Resilient Farming Systems in a Changing Climate, pp. 214–245. CAB International, Wallingford (2016)

18. Swanepoel, C.M., Swanepoel, L.H., Smith, H.J.: A review of conservation agriculture research in South Africa. South African J. Plant Soil **35**(4), 297–306 (2018)

19. Van Bruggen, A., He, M.M., Shin, K.: Environmental and health effects of the herbicide glyphosate. Sci. Total Environ. **616–617**, 255–268 (2018)

20. Yu, T., van Biljon, J., Marenya, M.O., Bobobee, E.Y.H., du Plessis, H.L.M.: Evaluation of soil compaction under conservation tillage in South Africa. In: 18th World Congress of CIGR, Beijing, China (2014)

Chapter 12
Ecological Significance of Seaweed Biomass Production and Its Role in Sustainable Agriculture

V. Veeragurunathan, K. G. Vijay Anand, Arup Ghosh, U. Gurumoorthy, and P. Gwen Grace

Introduction

Seaweeds are marine macroalgae, primitive, non-flowering photosynthetic plants and lack steam, root and leaf and occurring in intertidal, tidal, and subtidal regions of seas and oceans. They are natural renewable resources. Based on pigments and biochemical composition, seaweeds are classified into three major group of seaweeds mainly green (Chlorophyta), brown (Phaeophyta), and red (Rhodophyta) and grow abundantly along the coastline, particularly in the intertidal region due to the availability of the rocky substratum [103]. Seaweeds act as ecosystem engineer which providing space for many marine organisms and structuring and maintaining the coastal biodiversity of sea by providing space for marine microorganisms and higher organisms, as a nursery ground for fishes [82]. Seaweeds are also providing nutrition and shelter to diverse groups of invertebrates such as polychaetes, amphipods, isopods, and gastropods [10] which commonly observed on the surface of seaweeds and these organisms may form an important food source for juvenile fishes, which are also abundant in seaweed habitats [45]. Seaweeds are also considered as major primary producers in the marine ecosystems with great ecological and economic

V. Veeragurunathan (✉) · U. Gurumoorthy · P. Gwen Grace
CSIR-CSMCRI-Marine Algal Research Station, Mandapam Camp 623519, India
e-mail: veeragurunathan@csmcri.res.in

K. G. Vijay Anand · A. Ghosh
CSIR-Central Salt and Marine Chemicals Research Institute, Bhavnagar, Gujarat 364002, India
e-mail: kgvijay@csmcri.res.in

A. Ghosh
e-mail: arupghosh@csmcri.res.in

V. Veeragurunathan · A. Ghosh
Academy of Scientific and Innovative Research (AcSIR), Uttar Pradesh, Ghaziabad 201002, India

© Centre for Science and Technology of the Non-aligned and Other Developing Countries 2023
K. Pakeerathan (ed.), *Smart Agriculture for Developing Nations*, Advanced Technologies and Societal Change, https://doi.org/10.1007/978-981-19-8738-0_12

significance [28]. Seaweeds provide the basic direct food for the herbivores and indirectly to the carnivores and hence are a part of food web in marine ecosystem. 90% of the species of marine plants are algae and roughly 50% of the global photosynthesis from seaweeds [43]. At global level, 8950 species reported [11] and in India, 844 seaweed species are reported [69]. In 2018, 32.4 million tons of algae produced which included both wild-collected and cultivated biomass and seaweed farming contributed about 97.1% of total aquatic algal production [35]. The world production of seaweeds has increased three-fold in 2018, i.e., from 10.6 million tons in 2000 to 32.4 million tons in 2018. In the global scenario, about 221 species are having commercial utility and only 10 species are being commercially cultivated and had a market value of 11.7 billion US$. Among the 10 species, *Eucheuma* sp. (35%), *L. japonica* (27%), *Gracilaria* sp. (13%), *U. pinnadifida* (8%), *Kappaphycus alvarezii* (6%), *Porphyra* sp. (4%), have a major share in global seaweed biomass production [35]. *K. alvarezii* production in the world was 183,000 tons (dry) in 2010 while it was 1490 tons (fresh) during the same period in India. Seaweeds are the only resources for commercially important phycocolloids such as agar, carrageenan, and alginic acid production which have critical role in food, textile and leather industries as thickening and stabilizing agent [73]. Seaweed's phycocolloids production in 2015 was 93,035 tons wt and had a market value of 1058 million US$ [73]. Seaweeds are also used as human food, animal fodder, aqua feed, manure, and liquid seaweed fertilizer for crops [94]. Seaweeds contain considerable amount of micronutrients and plant growth hormones such as auxins, gibberellins, and cytokinins resulted promoting plant growth and seed germination [114]. Seaweeds are also used as immune boosters of plants and used in disease management [95]. Seaweeds have the capacity to take away dyes which is the effluent from various industries exclusively from printing, textile and paper industries, nutrients especially nitrogen and phosphorous, heavy metals, and phenolic compounds [8, 105] efficiently. The United Nations, through "The 2030 Agenda for Sustainable Development" identified 17 Sustainable Development Goals (SDGs) as a blueprint for addressing the present and future challenges of peace and prosperity facing the planet and mankind. Among the 17 goals, 'zero hunger', 'responsible consumption and production' as well as 'climate action' are more pertinent to the agriculture, forestry and land use (AFOLU) sector. It is well known that this sector contributes to less than quarter of the total anthropogenic GHG emissions [90]. Further, deforestation, nutrient management as well as agricultural emissions from livestock are perceived as the main GHG contributors in this sector. Thus, the focus has predominantly shifted to address these challenges. Among the several strategic themes such as conservation agriculture, integrated plant nutrient management, integrated pest management, management of resources employed for sustainable agricultural production, it is now widely recognized that biostimulants can play a significant role in crop intensification with limited environmental trade-offs [88]. Biostimulants include diverse group of substances as well as micro-organisms that can enhance plant growth when applied in low quantities [15, 26]. In this article, significant ecological roles of seaweeds such as carbon sequestration, biosorption of heavy metals, bio-accumulation of nutrients and heavy metals, biological indicators, bioremediation, food for fishes, and human, animal feed were described and also seaweeds role in sustainable agriculture was discussed.

Ecological Significance of Seaweeds

Carbon Sequestration

Massive use of fossil fuels has resulted in increase of the emission of greenhouse gas, carbon dioxide, in large quantities which have been trapped in the atmosphere and CO_2 emission is progressively increasing every year. According to an assessment of the Intergovernmental Panel on Climate Change (IPCC), it may reach to 20 billion tons/year by 2100 with disastrous environmental consequences. One of the main strategies being proposed to minimize concentrations of CO_2 in the earth's atmosphere is to increase the potential of carbon sequestration from the atmosphere by algae through photosynthetic processes since the marine ecosystem is one of the reservoirs that potentially absorb carbon emissions in large quantities and can capture up to 2.3 Gt C/year of total 6.3 Gt carbon released from anthropogenic activities [39]. Blue carbon-a new concept that declared as the carbon stored in marine and coastal ecosystems in biomass, buried in sediments, and sequestered from the atmosphere and ocean. Seaweeds could sequester higher amount of carbon than terrestrial plants due to its high adaptability. Seaweeds can convert inorganic carbon into carbohydrate through photosynthesis and permanently store it into their biomass and might be exported through food chain. Seaweed production, both from wild stocks and from aquaculture, represents an important channel for CO_2 removal from the atmosphere and can absorb 1.5 Pg C year^{-1} via their net production [53]. Seaweed aquaculture can be calculated at 2.48 million tons of CO_2 (0.68 Tg C) year^{-1} [27]. Erlania and Radiatra [31] stated that cultured species of *K. alvarezii* could absorb 27.35–158.40 ton CO_2/ha/year. Mashoreng et al. [62] reported that cultivation of *K. alvarezii* and *Eucheuma spinosum* at South Sulawesi showed carbon sequestration potential 57.64 ton CO_2/ha/yr. Fakhraini et al. [33] reported the carbon sequestration potential of *Kappaphycus striatum* as 66.07–125.51 tons C/ha/year.

It is estimated that the seaweed biomass along the Indian coast is capable of utilizing 9052 t CO_2/d with a net carbon credit of 8687 t/d and emission of 365 t CO_2/d [46]. The green seaweeds showed dominant in their carbon assimilation efficiency [46] than red and brown seaweeds. However, farmed seaweed showed higher carbon sequestration rate than natural seaweed resources [33]. In India, *K. alvarezii* is the only red alga cultivated commercially in states of Tamil Nadu and Gujarat and has proved to be a profitable venture along the pockets of Tamilnadu coastal region since 2003 [59]. In a hectare level cultivation of *K. alvarezii* (Fig. 12.1) can sequester 643.80 tons CO_2/ha/yr (Table 12.1). Similarly, in a hectare level cultivation of *Gracilaria debilis* (Fig. 12.2) and *Gracilaria edulis* (Fig. 12.3) can sequester 10.71 tons CO_2/ha/yr (Table 12.1).

Fig. 12.1 Raft culture of *K. alvarezii* (carrageenan yielding seaweed)

Biosorption

Effluents originating from many industries contain dissolved heavy metals. If these effluents are discharged without treatment, they will have adverse effects on the environment. Numerous techniques such as ion exchange, chemical precipitation, electrochemical methods, and membrane technology have been proposed by various researchers for the treatment of heavy metal-bearing effluents [84]. In recent years, biosorption, a popular method is considered in greater momentum for treating the heavy metal effluents. Seaweeds are excellent biosorbent for the removal of heavy metal ions from aqueous solutions, because of the presence of macromolecules [12, 84].

Seaweed biomass can be transformed into another sorbent biochar. Biochar is known as "Biological Charcoal" and prepared by pyrolysis process of the seaweed biomass [65, 79]. As an effective bio-adsorbent, seaweed biochar can be effectively used in wastewater treatment systems and have significant role in resource recycling [72]. Roberts et al. [79] stated that while treating biochar with waste, many contamination problems can be effectively fixed. Biomass of *Cladophora* spp., *Cladophoropsis* sp., *Chaetomorpha linum*, *Gracilaria* sp., *Saccharina japonica*, *Sargassum fusiforme*, *Ulva flexuosa*, *Undaria pinnatifida* used for biochar production [13, 19, 51, 79]. Biochar of *S. japonica* which obtained at higher than 400 °C

Table 12.1 Carbon sequestration rate of commercially valued seaweed's farming in hectare level

Name of the algae	Net biomass/raft (kg fr wt)	No of growth cycle/year	Available net biomass/raft/year (ton fr wt)	No of rafts/hectare	Net biomass/hectare/cycle (tons fr wt)	Net biomass/hectare/year (tons fr wt)	Carbon sequestration rate (kg CO_2/ton fr wt/year/ha)	Reference	Carbon sequestration rate (ton CO_2/ha/yr)
K. alvarezii	200	5	1	1500 (3 × 3 m)	300	1500	429.24	Mashoreng et al. [62]	643.86
G. debilis, G. edulis	25	6	0.150	2000 (2 × 2 m)	50	300	35.7	Zhou et al. [116]	10.71

Fig. 12.2 Raft culture of *G. debilis* (agar-yielding seaweed)

Fig. 12.3 Raft culture of *G. edulis* (agar yielding seaweed)

temperature showed greater efficiency (98%) in removal of cadmium, copper, and zinc ions [72]. Similarly, *S. fusiforme biochar* showed 86% of removal efficiencies for cadmium, copper, and zinc ions. Seaweed biochar can also remove oxy anionic elements such as As, Mo, and Se [44, 50]. Biochar produced from *K. alvarezii* exhibited three times more dye sorption capacities than the value obtained from raw biomass [98]. Biochar of *Turbinaria conoides* recorded the biosorption of dye Remazol brilliant blue R [105]. Seaweed biochar is rich in calcium, magnesium, sodium, and potassium which helped in removal of heavy metal ions due to their greater cation capacity exchange [22]. Another advantage of seaweed biochar is, after washing with deionized water, it can be reused for several times [91].

Bioremediation

Industrialization and urbanization largely contribute to the degradation of the natural environment. Removal of toxic metals especially arsenic, cadmium, mercury, lead, and organic compounds from the environment is highly essential. Heavy metals are non-degradable toxic pollutant and have long time persistence in the environment and can accumulate in the food chain through bio-accumulation process [102]. Bioremediation is the process to reduce the concentrations or toxic effects of contaminants in the aquatic ecosystems by using plants. Seaweeds are promising candidates for the remediation technologies [23]. Feed and fertilizers which are used for intensive aquaculture of fish, shrimp, and abalone and decomposition of excreta matter released from fish, shrimp, and abalone generated nitrogen and phosphorus-enriched nutrients in seawater which resulted eutrophicated seawater which pollute the coastal water in farming area and toxic to native fish and shrimp species [17, 25, 55, 67, 112, 113]. Therefore, combined seaweed cultivation in IMTA system is proposed to remove nutrients efficiently [61]. They reported that *Gracilaria caudata* which showed nutrient removal of 100% of NO_2; 72.4% of NO_3%; 29.8% of NH_4 and 44.5% dissolved inorganic nitrogen of seawater collected from shrimp farm. Elizondo-González et al. [30] reported that green macroalga *Ulva lactuca* recorded removal of 80% of nitrogen; 64% of phosphorus nutrients from wastewater coming from integrated shrimp's culture. In the IMTA system, *G. edulis* and *U. lactuca* removed significant 70% and 45%, ammonium in shrimp wastewater which helped in improving water quality and shrimp growth and survival [55].

Seaweed cultivation can be used for bioremediation of coastal eutrophicated waters since it act as a nutrient scrubber system [18, 80]. China's large-scale seaweed aquaculture contributed the share of 66% of the global seaweed production [109, 115] and by culturing seaweeds in eutrophicated waters in coastal region of China, 9500 tons of phosphorus and 75,000 tons of nitrogen can be removed annually [109]. Large-scale cultivation of *Porphyra yezoensis* in open sea in China to reduce the nutrient concentrations of seawater at 44% for NH_4^+N, 49% for NO_3^+N, and 45% for PO_4^+P and nutrient concentrations of seawater in the cultivation area were significantly lower than in the non-cultivation [108]. Cultivation of red seaweed *Gracilariopsis longissima* in Hangzhou Bay (East China Sea) reported that the nutrient removal at 54% for NH_4^+N, 75% for NO_3^+N, 49% for PO_4^+P [42].

Biological Indicators

The coastal environments throughout the world are experiencing constant exposure to pollutants from anthropogenic sources. A bioindicator is an organism which gives overall information on the presence or absence of a pollutant and the concentration. Many marine sessile organisms are considered as bioindicators for environmental quality assessment of the marine waters. Seaweeds are used as bioindicators because of their distribution, size, longevity, easy sampling, minimal impact on resident biota during sampling, presence at pollution sites, ability to accumulate metals, rapid reproduction rates, and very short life cycles, easy to culture in the laboratory [21, 106]. The seaweed species can directly reflect the water quality assessment because they are sensitive to some pollutants, and algal metabolism is sensitive to the variation of environmental and natural disturbances [106]. In order to detect the macro and microelements in the environment, seaweed are used for biomonitoring studies which mostly involved for inorganic contaminants (97%) and very few organic contaminants [36]. *Ulva lactuca* (green alga) reported as good biomonitor of toxic metal ions exclusively for cadmium, mercury and lead [41, 70]. Leal et al. [57] suggested that *Enteromorpha* sp. and *Porphyra* sp. can be used as biomonitors of Cd, Cu, Hg, and Pb. Al-Homaidan et al. [3] reported that green algae *Enteromorpha* and *Cladophora* have been utilized to measure heavy metal levels in many parts of the world. Akcali and Kucuksezgin [1] stated that brown alga *Cystoseira* sp., the green algae *Ulva* sp. and *Enteromorpha* sp. are cosmopolitan biomonitors for trace metals in the Aegean Sea. Chakraborty et al. [16] highlighted the use of green algae *U. lactuca* and *Enteromorpha intestinalis* and brown algae *Padina gymnospora* and *Dictyota bartayresiana* as biomonitors because of their high uptake of heavy metals. Wong et al. [107] used *Enteromorpha crinita* as a biomonitor in the Hong Kong waters, and Say et al. [83] recommended the use of *Enteromorpha* species as biomonitors in temperate region.

Animal Feed

The availability of seaweed for animals has been increased with the production of seaweed-based animal feed in the form of dried seaweed powder. Norway was among the early producers of seaweed-based animal feed using brown alga *Ascophyllum nodosum*. France has used *Laminaria digitata* for animal feed. Iceland has used both *Ascophyllum* and *Laminaria* species for animal feed. In Australia, the brown seaweed *Macrocystis pyrifera* and the red seaweed *G. edulis* have been used as feed for abalone. *Palmaria mollis*, *Haliotis rufescens*, and *U. lactuca* are used as a fish feed [97]. *Pelvetia* sp., *Rhodymenia palmata*, *Alaria* sp., *Fucus* sp., *Chondra filus*,

Ascophyllum sp., *Macrocystis* sp., *Palmaria* sp., and *Laminaria* sp. are the major genera of seaweeds used as fodder in various countries [14, 94]. Tocopheral and Vitamin E in seaweeds increased the fertility rate and birth rate of animals. Cattle fed with *Laminaria* have attained more natural resistance to diseases such as foot and mouth. Animal feed prepared from *K. alvarezii* and *Gracilaria heterocladia* in dry powdered form showed highest survival rate in prawn *Penaeus monodon* [52]. Several macroalgal genera such as *Ulva, Undaria, Ascophyllum, Porphyra, Sargassum, Poly-cavernosa, Gracilaria,* and *Laminaria* are widely used in fish diets [66, 94]. Feed additives for laying hens and pigs were prepared from mixture of two green seaweeds, namely, *Ulva prolifera* and *Cladophora* sp. with enriched microelement ions Cu (II), Zn(II), Co(II), Mn(II), Cr(III). These microelements enriched feed additive resulted in increased egg weight, eggshell thickness, and body weight of hens and enhanced yolk color in hen's eggs, and higher microelement transfer in meat of pigs [64].

Food

Many seaweed species are used as food throughout the Asia and the Pacific region [11]. Seaweeds have been traditionally used as food for human and animals. Seaweeds are rich sources of bioactive compounds such as carotenoids, dietary fiber, protein, essential fatty acids, vitamins, and minerals. *Laminaria* sp. (kombu), *Undaria* sp. (wakame), *Hizikia fusiforme* (hiziki), *Porphyra* (Nori), *Gracilaria coronopifolia* (Ogo Kim Chee, Sunomono), *Codium edule* (Gulamon sald), *Sargassum* sp. (horse-tail), *Rosenvingea* sp. (slippery cushion), *Turbinaria* sp. (spiny leaf), *Colpomenia* sp. (paperly sea bubble), *Hydroclatharus* sp. (sea colander), *Padina* sp. (sea fan ribbon) are edible and important food resource in the Asian countries especially China, Korea, and Japan. They are consumed raw, boiled, or dried material with sweetened green beans, jelly, crushed ice, and coconut milk in Southern Vietnam [11, 34, 101]. As the seaweed has high protein content, it is being used by many of the countries like Japan, China, Korea, Malaysia, Thailand, Indonesia, Philippines, and other South East Asia. Seaweeds like *Ulva* sp., *Enteromorpha* sp., *Caulerpa* sp., *Codium* sp., *Monostroma* sp., *Sargassum* sp., *Hydroclathrus* sp., *Laminaria* sp., *Undaria* sp., *Macrocystis* sp., *Porphyra* sp., *Gracilaria* sp., *Eucheuma* sp., *Laurencia* sp. and *Acanthophora* sp. are used in the preparation of soup, salad and curry [11, 68, 71], salad vegetable or as garnishing fish [94]. Red macro-algae (*Gracilaria* spp.) are used as a fresh food in Hawaii. Species commonly marketed include *G. coronopifolia, G. parvispora, G. salicornia,* and *G. tikvahiae* [11].

Role in Sustainable Agriculture

Soil Amendment by Seaweed

Seaweeds are used as a manure to increase the soil fertility due to its good moisture-holding capacity, organic nature, biodegradable, non-toxic, non-polluting, and non-hazardous to human, animals, and birds [71]. In ancient times, seaweed was composed with organic sources like peat, straw, etc., or mixed with sand and soil to protect the plants under abiotic and biotic stress and increase plant resistance against pest and diseases. During 1951, in European countries, seaweeds were used directly or in composted form with farmyard manure as soil conditioner to increase crop productivity in coastal areas and for recovery of alkaline soils [76]. *Ascophyllum*, *Ecklonia*, and *Fucus* are the general species sold as soil additives and functioned as both fertilizer and soil conditioner. They have a suitable content of nitrogen and potassium but are much lower in phosphorus than traditional animal manures and the typical N:P:K ratios compared to chemical fertilizers [34]. The addition of seaweed compost as an organic amendment to a horticultural soil has shown an increase in yield and resistance to diseases on tomato plants [32].

Seaweed biochar can be used for the bioremediation of polluted land with degraded soils and the bioremediation included sequestration of carbon in soil and soil amelioration [79]. For soil amelioration purpose, nutrient-rich biochar (N ranged from 0.3 to 2.8%, P from 0.5 to 6.6 g kg^{-1}, and K from 5.1 to 119 g kg^{-1}) prepared from intensively cultivated seaweeds genera *Eucheuma*, *Gracilaria Kappapaphycus*, *Saccharina*, *Sargassum*, *Undaria* showed common characters such as relatively low C content and surface area, essential trace elements (N, P, and K) and exchangeable cations (particularly K). The pH of seaweed biochar found in the range of 7 (neutral) to 11 (alkaline) which allowed broad-spectrum applications in diverse soil types. Due to a high fixed C content with a mineral-rich substrate, biochar enhanced crop productivity [79].

Biochar prepared by co-pyrolysis of nutrients-rich *Enteromorpha prolifera* and lignocellulose-rich corn straw in the ratio of 1:1 and 7:3 showed larger surface area, low Na content, and slower nutrient release rate and higher water-soluble N/P content helped in improvement of the soil properties and enhancing the total N/P uptake which resulted in significant increase of the plant biomass of cherry tomato [96].

In order to alleviate the lethal effect of Hg on the growth of fenugreek (*Trigonella foenumgraecum*), dried green seaweed *Codium iyengrii* was added to soil [9]. At lower dose level, mobility of mercury was greatly reduced. Pilot-scale research on effective removal of DDT, an organic pesticide from the soil by using dried powder of *Ulva* sp., *Gelidium* sp. showed the enhancement of the biodegradation (80%) of DDT after 6 weeks of application in waterlogged soils [47].

Seaweeds as a Source of Biostimulants

It is well documented in the literature that seaweeds are a potential source of biostimulants [49]. Seaweeds such as *A. nodosum, Ecklonia maxima, Fucus* sps., *Laminaria* sp., *Sargassum* sp., *Durvillaea potatorum* especially belonging to the brown algal group (Ochrophyta) have been extensively used for this purpose. However, several other seaweeds such as *K. alvarezii, G. edulis, Caulerpa* sps., *Ulva* sps., belonging to phylogenetically diverse groups of algae have also been commercially exploited for biostimulant production [4]. The various commercial products available in markets worldwide have been well documented elsewhere [7, 20, 49, 85]. Some of the commercial seaweed-based biostimulants products available in India are given in Table 12.2.

Table 12.2 Some commercial seaweed-based biostimulants available in markets in India

Product name	Seaweed used	Formulations	Modes of application	Dosage per acre	Company
Biovita	*A. nodosum*	Granules Liquid	Soil Foliar	500–1000 mL[a]	Pi Industries
KdalPasi	*Sargassum* sp	Granules Gel Drip Foliar spray	Soil Soil/fertigation Drip Foliar	3 kg 3–10 kg 5–10 L 2–5 mL per liter of water	SNAP Alginates
Sagarika	Red and brown algae	Liquid Solid	Soil Foliar/fertigation	250 mL 8–10 kg	IFFCO
Agrogain	Proprietary	Liquid	Foliar Drenching		Sea6Energy
Agrofort			Foliar/fertigation		
Aquasap	*K. alvarezii*	Liquid	Foliar/fertigation Spray dried powder	200 g per liter of water	AquaAgri
Dhanzyme G	Proprietary	Granules Liquid	Soil Foliar	5 kg	Dhanukaagritech
Biozyme granules	Proprietary	Granulated	Soil	8–16 kg	Biostadt
Biozyme drip		Liquid	Drip		
Biozyme power +			Foliar	12 kg (in 2 doses)	
Biopearl	*A. nodosum*	Liquid	Foliar/fertigation		Atul Ltd

[a]Dosage per hectare

Methods of Extraction and Formulations of Seaweed Biostimulants

Although the use of seaweeds in agriculture dates back to the roman times [40] their use was accentuated with the development of liquefaction technologies in late 1950s. Earlier, seaweeds that were washed ashore by tides were harvested and used directly in the field. Nowadays, seaweed-based biostimulants are prepared in various ways. The extraction methods range from water extracts, alkaline hydrolysis, acid hydrolysis, micro-wave assisted, enzymatic, pressurized liquid to super critical fluid extraction. The merits and demerits of these extraction procedures on the composition of the seaweeds and the commercial products thus generated and marketed around the world have been described elsewhere [29, 86]. Different formulations of seaweed-based biostimulants have been employed for growth and yield enhancement either in a standalone manner or in combinations with other biostimulants such as humic and fulvic acids, hydrolysates (Dhanzyme Gold of Dhanukaagritech), amino acids, microorganisms [38], potassium (Tarma of Sea6energy), or their combinations (Max Grow, Amaze- of Biostadt). It has been generally agreed upon, that there exists a synergistic effect between the various biostimulants while promoting plant growth [81]. One can come across generally three broad categories of seaweed-based biostimulants formulations in the market. These include granular, gel based, and liquid formulations. Although the former two are generally used for soil-based applications, the latter can be used either for foliar applications or for fertigation. The percentage of the seaweed in these formulations may range from 20 to 80% with 20–28% w/v being the most common range. The solid/gel formulations can be applied either as a basal dose along with recommended rate of NPK fertilizers or at specific growth stages of the crop. Gel-based/liquid formulations can be also applied along with irrigation water and are advantageous as they can help reduce additional labor costs involved in cultivation (see Table 12.2 for the products). Compatibility of seaweed formulations with fungicides/insecticides would also be an added advantage; however, there is no empirical evidence on benefits of this. The rate of application or the dosages of the seaweed-based biostimulants tend to vary with the product as given in Table 12.2. This depends on the amount of seaweed concentrate present in the product, the type of biostimulant formulation, spray volume in case of liquid formulations, the plant density, crop used, canopy architecture, etc. Hence, one has to rely on the recommendation of the dosages given in the product for optimum results. However, in case of empirical studies, it was observed that prisitine extracts of seaweed biostimulants were evaluated in the range of 2.5–15% in case of *K. alvarezii, Sargassum* spp., *G. edulis* extracts [89, 99, 100, 104].

Active Principles in Seaweed Based Biostimulants

Since seaweed-based biostimulants are applied in small quantities, it is generally perceived that the active constituents are plant growth regulating substances such as phytohormones. The presence of these growth regulators have been identified in several seaweed-based biostimulants [20, 63, 93]. In addition, several other active principles have also been identified ranging from quaternary ammonium compounds like betaines [63], phenolic compounds [78]; various carbohydrates ranging from simple sugars like mannitol [58], to oligosaccharides of carrageenan [87] and finally polymers like alginic acid, fucoidan, laminarin, etc. [92]. The modes of action and growth stimulatory factor in seaweed biostimulants have been reviewed by [49] as well as [4]. These active principles are thought to function either alone or synergistically manner during plant response. There is a consensus among the scientific community that these constituents play important roles in enhancing the metabolic or physiological processes that ultimately lead to improvement in crop growth, better response to biotic and abiotic stresses.

Benefits of Seaweed-Based Biostimulants on Crop Growth, Yield, and Quality of Produce

A glimpse of the literature reveals that seaweed-based biostimulants have been employed for enhancing growth and yield of many crops [4, 15, 85]. For instance, efficacy of extracts of *K. alvarezii*, *G. edulis* have been tested on various crops such as maize [89], rice [56], pulses [74, 77], potato [75], sugarcane [48, 88].

The increase in yield in these cases was in range of 8–25% over control. Similarly, biostimulants obtained from *A. nodosum* have on a number of horticultural crops such as strawberries [60], carrots [2], spinach [111], cabbage [110], tomato [24]. Various quality parameters such as antioxidant content of the produce, the fiber and carbohydrate, and nutrient contents were enhanced due to application of seaweed biostimulants.

Benefits of Seaweed-Based Biostimulants on Soil Parameters and Environmental Impacts

The most important determining factor in employing seaweed-based biostimulants in agriculture is their ability to enhance yield of crops with limited or minimal detrimental effects to environment. One of the prerequisites of a sustainable process is that the production chain has to be environmentally benign. In this context, [5, 37] have reported that the life cycle assessment of biostimulants produced from *K. alvarezii* as well as *G. edulis* had a low carbon foot print, i.e., 118.6 and 73.1 kg CO_2 equivalents, respectively, per kiloliter of biostimulants production at factory gate. In addition, benefits in 18 different environmental impact categories were also reported like climate change which is similar to Greenhouse warming potential, human toxicity, marine, and fresh water eutrophication, terrestrial ecotoxicity, ozone depletion, etc. In contrast, production of synthetic fertilizers especially that of Nitrogen fertilizers is highly energy intensive and is environmentally detrimental. Further, the use phase of nitrogen fertilizers results in nitrogen dioxide emissions due to leaching and volatilization accentuating the impacts. In contrast, seaweed-based biostimulants have a low N content (> 500 mg L^{-1}) which may be directly absorbed by plants in foliar applications [88] or may be subject to limited microbial activity when applied to soil. The environmental benefits of the use of seaweed-based biostimulants have been reviewed by Anand and Ghosh [6]. The benefits especially in term of climate change owing to the use of seaweed biostimulants in rice, maize, as well as sugarcane crops was discussed. These benefits ranged from 2.2 to 9.5 kg CO_2 equivalent per ton of produce. Although application of biostimulants to soil especially of those containing alginic acid or other polysaccharides often results in increased porosity and changes in bulk density of the soil resulting in increased aeration as well as altering the water retention. Further, it was observed that this is a diversity shift in microbes. Thus, they can act as soil conditioners enabling the plants to overcome both biotic and abiotic factors that influence the rhizosphere and its vicinity. In addition, one important feature that was observed in seaweed biosimulant application was increased nutrient use efficiency, especially with respect to N uptake. This would have significant consequences as less of applied N would be subject of volatilization and microbial degradation leading to lower emissions [88]. Transcriptomic studies in maize by Kumar et al. [54] have shown that indeed the biostimulants applied increased the transcript levels of enzymes involved in N transport and uptake in maize roots thus substantiating the above claims.

Conclusion

Biomass produced from seaweeds cultivation act as raw material for phycocolloid and bio-stimulant production and as well as key role in providing employment opportunity to coastal rural population and improved their livelihood. Apart from these, seaweeds also play important role in maintaining healthy nature of marine environment by bioremediation and biosorption process. Seaweeds which are used in the bioremediation could be a significant source of biomass for several valuable products and also for biofuels, biosorbents, and animals feed additives. It is evident that seaweed-based biostimulants not only enhance growth and productivity but also enhance the quality of produce in an environmentally sustainable fashion. Thus, it is high time that policy decision be made for using these biostimulants as an alternative practice/strategy for improving crop productivity.

Acknowledgements The authors expressed their sincere thanks to Dr. S. Kannan, CSMCRI, Bhavnagar, Gujarat for his encouragement to pursue the bio-stimulant works and as well as for allowing us to write this article. The authors expressed their sincere to NAM –S&T centre, New Delhi for providing opportunity to submit this article as conference proceeding paper. Financial support from SERB-DST, New Delhi **(CRG/2019/005304)** and CSIR, New Delhi is greatly acknowledged. This contribution has CSIR-CSMCRI PRIS registration number CSIR-CSMCRI 223/2021.

References

1. Akcali, I., Kucuksezgin, F.: A biomonitoring study: heavy metals in macroalgae from eastern Aegean coastal areas. Mar. Pollut. Bull. **62**, 637–645 (2011). https://doi.org/10.1016/j.marpol bul.2010.12.021
2. Alam, M.Z., Braun, G., Norrie, J., Mark Hodges, D.: *Ascophyllum* extract application can promote plant growth and root yield in carrot associated with increased root-zone soil microbial activity. Can. J. Plant Sci. **94**(2), 337–348 (2014). https://doi.org/10.4141/cjps20 13-135
3. Al-Homaidan, A.A., Al-Ghanayem, A.A., Alkhalifa, A.H.: Green algae as bioindicators of heavy metal pollution in Wadi Hanifah Stream, Riyadh, Saudi Arabia. Int. J. Water Res. Arid Environ. **1**, 10–15 (2011)
4. Ali, O., Ramsubhag, A., Jayaraman, J.: Biostimulant properties of seaweed extracts in plants: implications towards sustainable crop production. Plants **10**(3), 531 (2021). https://doi.org/ 10.3390/plants10030531
5. Anand, K.V., Eswaran, K., Ghosh, A.: Life cycle impact assessment of a seaweed product obtained from *Gracilaria edulis*—a potent plant biostimulant. J. Cleaner Prod. **170**, 1621–1627 (2018). https://doi.org/10.1016/j.jclepro.2017.09.241
6. Anand, K.G.V., Ghosh, A.: Seaweed biostimulants for climate change adaptations in dryland agriculture in semi-arid areas. In: Poshiwa, X., Ravindra Chary, G. (eds.) Climate Change Adaptations in Semi-Arid Areas. (in proof) (2021)
7. Arioli, T., Mattner, S.W., Winberg, P.C.: Applications of seaweed extracts in Australian agriculture: past, present and future. J. Appl. Phycol. **27**(5), 2007–2015 (2015). https://doi.org/ 10.1007/s10811-015-0574-9
8. Arumugam, N., Chelliapan, S., Kamyab, H., Thirugnana, S., Othman, N., Nasri, N.S.: Treatment of wastewater using seaweed: a review. Int. J. Environ. Res. Public Health **15**, 2851 (2018). https://doi.org/10.3390/ijerph15122851
9. Azmat, R., Askari, S.: Improvement in the bioenergetics system of plants under Hg stress environment via seaweeds. Pak. J. Bot. **47**(3), 851–858 (2015)
10. Ba-akdah, M.A., Satheesh, S., Al-sofyani, A.A.: Habitat preference and seasonal variability of epifaunal assemblages associated with macroalgal beds on the Central Red Sea coast, Saudi Arabia. J. Marine Biol. Assoc. UK **96**(7), 1457–1467 (2016). https://doi.org/10.1017/S00253 15415001678
11. Berna, K., Semra, C., Gamze, T., Hatice, T., Edis, K.: Seaweeds for food and industrial applications. Food Ind. Innocenzo Muzzalupo Intech Open (2013).https://doi.org/10.5772/ 53172
12. Bilal, M., Rasheed, T., Sosa-Hernandez, J.E., Raza, A., Nabeel, F., Iqbal, H.M.N.: Biosorption: an interplay between marine algae and potentially toxic elements a review. Marine Drugs **16**, 65 (2018). https://doi.org/10.3390/md16020065
13. Bird, M.I., Wurster, C.M., Silva, P.H.D., Paul, N.A., de Nys, R.: Algal biochar: effects and applications. Glob. Change Biol. Bioenergy **4**, 61–69 (2012). https://doi.org/10.1111/j.1757-1707.2011.01109.x
14. Boney, E.: Aspects of the biology of the seaweeds of economic importance. Adv. Mar. Biol. **3**, 105–253 (1965). https://doi.org/10.1016/S0065-2881%2808%2960397-1
15. Calvo, P., Nelson, L., Kloepper, J.W.: Agricultural uses of plant biostimulants. Plant Soil **383**(1), 3–41 (2014). https://doi.org/10.1007/s11104-014-2131-8
16. Chakraborty, S., Bhattacharya, T., Singh, G., Maity, J.P.: Benthic macroalgae as biological indicators of heavy metal pollution in the marine environments: a biomonitoring approach for pollution assessment. Ecotoxicol. Environ. Saf. **100**, 61–68 (2014). https://doi.org/10.1016/j. ecoenv.2013.12.003
17. Chopin, T.: Aquaculture, integrated multi-trophic (IMTA). In: Christou, P., Savin, R., Costa-Pierce, B.A., Misztal, I., Whitelaw, C.B.A. (eds.) Sustainable Food Production. Springer, New York

18. Chung, I.K., Kang, Y.H., Yarish, C., Kraemer, G.P., Lee, J.A.: Application of seaweed cultivation to the bioremediation of nutrient-rich effluent. Algae **17**(3), 187–194 (2002). https://doi.org/10.4490/ALGAE.2002.17.3.187

19. Cho, H.J., Baek, K., Jeon, J.K., Park, S.H., Suh, D.J., Park, Y.K.: Removal characteristics of copper by marine macroalgae-derived chars. Chem. Eng. J. **217**, 205–211 (2013). https://doi.org/10.1016/j.cej.2012.11.123

20. Crouch, I.J., Van Staden, J.: Evidence for the presence of plant growth regulators in commercial seaweed products. Plant Growth Regul. **13**(1), 21–29 (1993). https://doi.org/10.1007/BF00207588

21. Cruz, C., Varma, A.: Utilization of seaweed in soil fertilization-salt tolerance. In: Nabti, E. (ed) Biotechnological Applications of Seaweeds. Nova Science Publishers, Inc., pp. 16–24 (2017)

22. Davis, T.A., Volesky, B., Mucci, A.: A review of the biochemistry of heavy metal biosorption by brown algae. Water Res. **37**, 4311–4330 (2003).

23. Dawes, C.: Macroalgae systematics. In: Fleurence, J., Levine, I. (eds.) Seaweed in Health and Disease Prevention. Amsterdam: Elsevier Inc., p. 107e138 (2016)

24. Di Stasio, E., Van Oosten, M.J., Silletti, S., Raimondi, G., dell'Aversana, E., Carillo, P., Maggio, A.: *Ascophyllum nodosum*-based algal extracts act as enhancers of growth, fruit quality, and adaptation to stress in salinized tomato plants. J. Appl. Phycol. **30**(4), 2675–2686 (2018). https://doi.org/10.1007/s10811-018-1439-9

25. Dosdat, A., Servais, F., Métailler, R., Huelvan, C., Desbruyeres, E.: Comparison of nitrogenous losses in five teleost fish species. Aquaculture **141**, 107–127 (1996). https://doi.org/10.1016/0044-8486(95)01209-5

26. Du Jardin, P.: Plant biostimulants: Definition, concept, main categories and regulation. Sci. Hortic. **196**, 3–14 (2015). https://doi.org/10.1016/j.scienta.2015.09.021

27. Duarte, C.M., Cebrian, J.: The fate of marine autotrophic production. Limnol. Oceanogr. **41**, 1758–1766 (1996). https://doi.org/10.4319/lo.1996.41.8.1758

28. Egan, S., Harder, T., Burke, C., Steinberg P., Kjelleberg, S., Thomas, T.: The seaweed holobiont: understanding seaweed-bacteria interactions. FEMS Microbiol. Rev. **37**, 462–476 (2013). https://doi.org/10.1111/1574-6976.12011

29. El Boukhari, M. E., Barakate, M., Bouhia, Y., and Lyamlouli, K., 2020. Trends in seaweed extract based biostimulants: manufacturing process and beneficial effect on soil-plant systems. Plants, 9(3), 359.https://doi.org/10.3390/plants9030359

30. Elizondo-González, R., Quiroz-Guzmán, E., Escobedo-Fregoso, C., Magallón-Servín, P., Peña-Rodríguez, A.: Use of seaweed Ulva lactuca for water bioremediation and as feed additive for white shrimp *Litopenaeus vannamei*. Peer J. **6**, e4459. (2018) https://doi.org/10.7717/peerj.4459.10.7717/peerj.4459

31. Erlania, Radiarta, I.N.: The use of seaweeds aquaculture for carbon sequestration: a strategy aquaculture for climate change mitigation. J. Geodesy Geomatics Eng. **2**, 109–115 (2015). https://doi.org/10.17265/2332-8223/2015.06.006

32. Eyrasa, M.C., Defosséb, G.E., Dellatorrea, F.: Seaweed compost as an amendment for horticultural soils in Patagonia, Argentina. Compost Sci. Utilization **16**(2), 119–124 (2008). https://doi.org/10.1080/1065657X.2008.10702366

33. Fakhraini, M.S., Wisnu, W., Khathir, R., Patria, M.P.: Carbon sequestration in macroalgae *Kappaphycus striatum* in seaweed aquaculture site Alaang village, Alor Island, East Nusa Tenggara. Earth Environ. Sci. **404**, 012044 (2019). https://doi.org/10.1088/1755-1315/404/1/012044

34. FAO: Aguide to the seaweed industry. FAO Fisheries Technical Paper 44, Rome (2003)

35. FAO: The global status of seaweed production, trade and utilization. Globefish Res. Progra. **24**, 1–210 (2018)

36. García-Seoane, R., Fernández, J., Villares, R., Aboal, J.: Use of macroalgae to biomonitor pollutants in coastal waters: Optimization of the methodology. Ecol. Ind. **84**, 710–726 (2018). https://doi.org/10.1016/j.ecolind.2017.09.015

37. Ghosh, A., Anand, K.V., Seth, A.: Life cycle impact assessment of seaweed based biostimulant production from onshore cultivated *Kappaphycus alvarezii* (Doty) Doty ex Silva—is it environmentally sustainable? Algal Res. **12**, 513–521 (2015). https://doi.org/10.1016/j.algal.2015.10.015

38. González-González, M.F., Ocampo-Alvarez, H., Santacruz-Ruvalcaba, F., Sánchez-Hernández, C.V., Casarrubias-Castillo, K., Becerril-Espinosa, A., Castañeda-Nava, J.J., Hernández-Herrera, R.M.: Physiological, ecological, and biochemical implications in tomato plants of two plant biostimulants: Arbuscular mycorrhizal fungi and seaweed extract. Front. Plant Sci. **11**, 999 (2020). https://doi.org/10.3389/fpls.2020.00999

39. Hairiah, K., Dewi, S., Agus, F., Velarde, S., Ekadinata, A., Rahayu, S., Noordwijk, M.V.: Measuring Carbon Stocks: Across Land Use Systems: A Manual. World Agroforestry Centre, Bogor (2011)

40. Henderson, J.: The Roman Book of Gardening. Routledge, London, 152 p. (2004) https://doi.org/10.1017/S0075435800002811

41. Henriques, B., Lopes, C.B., Figueira, P., Rocha, L.S., Duarte, A.C., Vale, C., Pardale, M.A., Pereiraa, E.: Bioaccumulation of Hg, Cd and Pb by *Fucus vesiculosus* in single and multimetal contamination scenarios and its effect on growth rate. Chemosphere **171**, 208–222 (2017). https://doi.org/10.1016/j.chemosphere.2016.12.086

42. Huo, Y.Z., Xu, S.N., Wang, Y.Y., Zhang, J.H., Zhang, Y.J., Wu, W.N.: Bioremediation efficiencies of *Gracilaria verrucosa* cultivated in an enclosed sea area of Hangzhou Bay, China. J. Appl. Phycol. **23**, 173–182 (2011). https://doi.org/10.1007/s10811-010-9584-9

43. John, D.M., Tittley, I., Lawson G.W., Pugh, P.J.A.: Distribution of seaweed floras in the Southern Ocean. Bot. Mar. **37**, 235–239 (1994). https://doi.org/10.1515/botm.1994.37.3.235

44. Johansson, C.L., Paul, N.A., de Nys, R., Roberts, D.A.: The complexity of biosorption treatments for oxyanions in a multi-element mine effluent. J. Environ. Manage. **151**, 386–392 (2016). https://doi.org/10.1016/j.jenvman.2014.11.031

45. Jones, G.P.: Ecology of rocky reef fish of north-eastern New Zealand a review. NZ J. Mar. Fresh **22**, 445–462 (1988)

46. Kaladharan, P., Veena, S., Vivekanandan, E.: Carbon sequestration by a few marine algae: observation and projection. J. Mar. Biol. Ass. India **51**(1), 107–110 (2009)

47. Kantachote, D., Naidu, R., Williams, B., McClure, N., Megharaj, M., Singleton, I.: Bioremediation of DDT-contaminated soil: enhancement by seaweed addition. J. Chem. Technol. Biotechnol. **79**, 632–663 (2004). https://doi.org/10.1002/jctb.1032

48. Karthikeyan, K., Shanmugam, M.: The effect of potassium-rich biostimulant from seaweed *Kappaphycus alvarezii* on yield and quality of cane and cane juice of sugarcane var. Co 86032 under plantation and ratoon crops. J. Appl. Phycol. **29**(6), 3245–3252 (2017). https://doi.org/10.1007/s10811-017-1211-6

49. Khan, W., Rayirath, U.P., Subramanian, S., Jithesh, M.N., Rayorath, P., Hodges, D.M., Prithiviraj, B.: Seaweed extracts as biostimulants of plant growth and development. J. Plant Growth Regul. **28**(4), 386–399 (2009). https://doi.org/10.1007/s00344-009-9103-x

50. Kidgell, J.T., de Nys, R., Hu, Y., Paul, N.A., Roberts, D.A.: Bioremediation of a complex industrial effluent by biosorbents derived from freshwater macroalgae. PLoS One **9**, e94706 (2014). https://doi.org/10.1371/journal.pone.0094706

51. Kim, B.K., Lee, H.W., Park, S.H., Baek, K., Jeon, J.K., Cho, H.J., Jung, S.C., Kim, S.C., Park, Y.K.: Removal of Cu2+ by biochars derived from green macroalgae. Environ. Sci. Pollut. Res. **23**, 985–994 (2016). https://doi.org/10.1007/s11356-015-4368-z

52. Kotiya, A.S., Gunalan, B., Jetani, K.J., Solanki, J.B., Kumaran, R.: Comparison of *Penaeus monodon* (Crustacea, Panaeidae) growth between commercial feed vs commercial shrimp feed supplemented with *Kappaphycus alvarezii* (Rhodophyta, Solieriaceae) seaweed sap. AACL Bioflux **4**(3), 292–300 (2011)

53. Krause-Jensen, D., Duarte, C.M.: Substantial role of macroalgae in marine carbon sequestration. Nat. Geosci. **9**, 737–742 (2016). https://doi.org/10.1038/NGEO2790.doi:10.1038/ngeo2790

54. Kumar, R., Trivedi, K., Anand, K.V., Ghosh, A.: Science behind biostimulant action of seaweed extract on growth and crop yield: insights into transcriptional changes in roots of maize treated with *Kappaphycus alvarezii* seaweed extract under soil moisture stressed conditions. J. Appl. Phycol. **32**(1), 599–613 (2020). https://doi.org/10.1007/s10811-019-019 38-y

55. Lavania-Baloo, I.N., Salihi, I.U., Zainoddin, J.: The use of macroalgae (*Gracilaria changii*) as bio-adsorbent for Copper (II) removal. IOP Conf. Ser. Mater. Sci. Eng. **2017**, 012031 (2017). https://doi.org/10.1088/1757-899X/201/1/012031

56. Layek, J., Das, A., Idapuganti, R.G., Sarkar, D., Ghosh, A., Zodape, S.T., Meena, R.S.: Seaweed extract as organic bio-stimulant improves productivity and quality of rice in eastern Himalayas. J. Appl. Phycol. **30**(1), 547–558 (2018). https://doi.org/10.1007/s10811-017-1225-0

57. Leal, M.F.C., Vasconcelos, M.T.V., Sousa-Pinto, I., Cabral, J.P.S.: Biomonitoring with benthic macroalgae and direct assay of heavy metals in seawater of the oporto coast (Northwest Portugal). Mar. Pollut. Bull. **34**, 1006–1015 (1997)

58. Lötze, E., Hoffman, E.W.: Nutrient composition and content of various biological active compounds of three South African-based commercial seaweed biostimulants. J. Appl. Phycol. **28**(2), 1379–1386 (2016). https://doi.org/10.1007/s10811-015-0644-z

59. Mantri, V.A., Eswaran, K., Shanmugam, M., Ganesan, M., Veeragurunathan, V., Thiruppathi, S., Reddy, C.R.K., Seth, A.: An appraisal on commercial farming of *Kappaphycus alvarezii* in India: success in diversification of livelihood and prospects. J. Appl. Phycol. **17**, 335–357 (2017). https://doi.org/10.1007/s10811-016-0948-7

60. Mattner, S.W., Milinkovic, M., Arioli, T.: Increased growth response of strawberry roots to a commercial extract from *Durvillaea potatorum* and *Ascophyllum nodosum*. J. Appl. Phycol. **30**(5), 2943–2951 (2018). https://doi.org/10.1007/s10811-017-1387-9

61. Marinho-Soriano, E., Azevedo, C.A.A., Trigueiro, T.G., Pereira, D.C., Carneiro, M.A.A., Camara, M.R.: Bioremediation of aquaculture wastewater using macroalgae and Artemia. Int. Biodeterior. Biodegradation **65**, 253–257 (2011). https://doi.org/10.1016/j.ibiod.2010.10.001

62. Mashoreng, S., La Nafie, Y.A., Isyrini, R.: Cultivated seaweed carbon sequestration capacity. Earth Environ. Sci. **370**, 012017 (2019). https://doi.org/10.1088/1755-1315/370/1/012017

63. Mondal, D., Ghosh, A., Prasad, K., Singh, S., Bhatt, N., Zodape, S.T., Ghosh, P.K.: Elimination of gibberellin from *Kappaphycus alvarezii* seaweed sap foliar spray enhances corn stover production without compromising the grain yield advantage. Plant Growth Regul. **75**(3), 657–666 (2015). https://doi.org/10.1007/s10725-014-9967-z

64. Michalak, I., Chojnacka, K., Dobrzanski, Z., Gorecki, H., Zielinska, A., Korczynski, M., Opaliński, S.: Effect of macroalgae enriched with microelements on egg quality parameters and mineral content of eggs, eggshell, blood, feathers and droppings. J. Anim. Physiol. Anim. Nutr. **95**, 374–387 (2011). https://doi.org/10.1111/j.1439-0396.2010.01065.x

65. Michalak, I.: Experimental processing of seaweeds for biofuels. WIRES **7**, e288 (2018). https://doi.org/10.1002/wene.288

66. Nakagawa, H., Montgomery, W.L.: Algae. In: Nakagawa, H., Sato, S., Gatlin III, D., (eds) Dietary Supplements for the Health and Quality of Cultured Fish. CABI North American Office Cambridge, MA02139, USA, pp. 133–168

67. Neori, A., Chopin, T., Troell, M., Buschmann, A.H., Kraemer, G.P., Halling, C.: Integrated aquaculture: Rationale, evolution and state of the art emphasizing seaweed biofiltration in modern mariculture. Aquaculture **231**(1e4), 361–391 (2004). https://doi.org/10.1016/j.aquaculture.2003.11.015

68. Novaczek, I.: A guide to the Common and Edible and Medicinal Sea Plants of the Pacific Island, University of the South Pacific, 40p (2001)

69. Oza, R.M., Zaidi, S.H.: A revised checklist of Indian marine algae. Central Salt & Marine Chemicals Research Institute, Bhavnagar, India p. 286 (2001)

70. Ozyigit, I., Uyanik, O.L., Sahin, N.R., Yalcin, I.E., Demir, G.: Monitoring the pollution level in Istanbul Coast of the Sea of Marmara using algal species *Ulva lactuca* L. Polish J. Environ. Stud. **26**(2) (2017). https://doi.org/10.15244/pjoes/66177

71. Pati, M.P., Sharma, S.D., Nayak, L., Panda, C.R.: Uses of seaweed and its application to human welfare: a review. Int. J. Pharm. Pharm. Sci. **8**(10), 12–20 (2016). https://doi.org/10.22159/ijpps.2016v8i10.12740

72. Poo, K.M., Son, E.B., Chang, J.S., Ren, X., Choi, Y.J., Chae, K.J.: Biochars derived from wasted marine macro-algae (*Saccharina japonica* and *Sargassum fusiforme*) and their potential for heavy metal removal in aqueous solution. J. Environ. Manag. **206**, 364–372 (2018). https://doi.org/10.1016/j.jenvman.2017.10.056

73. Porse, H., Rudolph, B.: The seaweed hydrocolloid industry: 2016 updates, requirements, and outlook. J. Appl. Phycol. **29**, 2187–2200 (2017). https://doi.org/10.1007/s10811-017-1144-0

74. Pramanick, B., Brahmachari, K., Ghosh, A.: Effect of seaweed saps on growth and yield improvement of green gram. Afr. J. Agric. Res. **8**(13), 1180–1186 (2013). https://doi.org/10.5897/AJAR12.1894

75. Pramanick, B., Brahmachari, K., Mahapatra, B.S., Ghosh, A., Ghosh, D., Kar, S.: Growth, yield and quality improvement of potato tubers through the application of seaweed sap derived from the marine alga *Kappaphycus alvarezii*. J. Appl. Phycol. **29**(6), 3253–3260 (2017). https://doi.org/10.1007/s10811-017-1189-0

76. Raghunandan, B.L., Vyas, R.V., Patel, H.K., Jhala Y.K.: Perspectives of Seaweed as Organic Fertilizer in Agriculture (2019). https://doi.org/10.1007/978-981-13-5904-0_13

77. Rathore, S.S., Chaudhary, D.R., Boricha, G.N., Ghosh, A., Bhatt, B.P., Zodape, S.T., Patolia, J.S.: Effect of seaweed extract on the growth, yield and nutrient uptake of soybean (*Glycine max*) under rainfed conditions. S. Afr. J. Bot. **75**(2), 351–355 (2009). https://doi.org/10.1016/j.sajb.2008.10.009

78. Rengasamy, K.R., Kulkarni, M.G., Stirk, W.A., Van Staden, J.: Eckol-a new plant growth stimulant from the brown seaweed *Ecklonia maxima*. J. Appl. Phycol. **27**(1), 581–587 (2015). https://doi.org/10.1007/s10811-014-0337-z

79. Roberts, D.A., Paul, N.A., Dworjanyn, S.A., Hu, Y., Bird, M.I., de Nys, R.: *Gracilaria* waste biomass (sampah rumput laut) as a bioresource for selenium biosorption. J. Appl. Phycol. **27**, 611–620 (2015)

80. Ross, M.E., Davis, K., McColl, R., Stanley, M.S., Day, J.G., Semi~ao, A.J.C.,: Nitrogen uptake by the macro-algae *Cladophora coelothrix* and *Cladophora parriaudii*: Influence on growth, nitrogen preference and biochemical composition. Algal Res. **30**, 1–10 (2018). https://doi.org/10.1016/j.algal.2017.12.005

81. Sandepogu, M., Shukla, P.S., Asiedu, S., Yurgel, S., Prithiviraj, B.: Combination of *Ascophyllum nodosum* extract and humic acid improve early growth and reduces post-harvest loss of Lettuce and Spinach. Agriculture **9**(11), 240 (2019). https://doi.org/10.3390/agriculture9110240

82. Satheesh, S., Wesley, S.G.: Diversity and distribution of macroalgae in the Kudankulam coastal waters, South-Eastern coast of India. Biodivers. J. **3**(1), 79–84 (2012)

83. Say, P.J., Burrows, I.G., Whitton, B.A.: Enteromorpha as a monitor of heavy metals in Estuarine and Coastal Intertidal waters. A method for the sampling, treatment and analysis of the seaweed *Enteromorpha* to monitor heavy metals in estuaries and coastal waters. Occasional Publications No. 2, Northern Environmental Consultants LH., Consett, Co., Durham (1986)

84. Senthilkumar, R., Vijayaraghavan, K., Thilakavathi, M., Iyer, P.V.R., Velan, M.: Seaweeds for the remediation of wastewaters contaminated with zinc(II) ions. J. Hazard. Mater. (2006). https://doi.org/10.1016/j.jhazmat.2006.01.014

85. Sharma, H.S., Fleming, C., Selby, C., Rao, J.R., Martin, T.: Plant biostimulants: a review on the processing of macroalgae and use of extracts for crop management to reduce abiotic and biotic stresses. J. Appl. Phycol. **26**(1), 465–490 (2014). https://doi.org/10.1007/s10811-013-0101-9

86. Shukla, P.S., Mantin, E.G., Adil, M., Bajpai, S., Critchley, A.T., Prithiviraj, B.: *Ascophyllum nodosum*-based biostimulants: sustainable applications in agriculture for the stimulation of plant growth, stress tolerance, and disease management. Front. Plant Sci. **10**, 655 (2019). https://doi.org/10.3389/fpls.2019.00655

87. Shukla, P.S., Borza, T., Critchley, A.T., Prithiviraj, B.: Carrageenans from red seaweeds as promoters of growth and elicitors of defense response in plants. Front. Marine Sci. **3**, 81 (2016). https://doi.org/10.3389/fmars.2016.00081

88. Singh, I., Anand, K.V., Solomon, S., Shukla, S.K., Rai, R., Zodape, S.T., Ghosh, A.: Can we not mitigate climate change using seaweed based biostimulant: a case study with sugarcane cultivation in India. J. Clean. Prod. **204**, 992–1003 (2018). https://doi.org/10.1016/j.jclepro. 2018.09.070

89. Singh, S., Singh, M.K., Pal, S.K., Trivedi, K., Yesuraj, D., Singh, C.S., Ghosh, A.: Sustainable enhancement in yield and quality of rain-fed maize through Gracilariaedulis and *Kappaphycus alvarezii* seaweed sap. J. Appl. Phycol. **28**(3), 2099–2112 (2016). https://doi.org/10.1007/s10 811-015-0680-8

90. Smith, P., Bustamante, M., Ahammad, H., Clark, H., Dong, E.A., Elsiddig, H., Haberl, R., Harper, J., House, M., Jafari, O., Masera, C., Mbow, N.H., Ravindranath, C.W., Rice, C., Robledo Abad, A., Romanovskaya, F., Sperling, Tubiello, F.: Agriculture, forestry and other land use (AFOLU). In: Edenhofer, O., Pichs-Madruga, R., Sokona, Y., Farahani, E., Kadner, S., Seyboth, K., Adler, A., Baum, I., Brunner, S., Eickemeier, P., Kriemann, B., Savolainen, J., Schlömer, S., von Stechow, C., Zwickel, T., Minx, J.C., (eds.) Climate Change 2014: Mitigation of Climate Change. Contribution of Working Group III to the Fifth Assessment Report of the Intergovernmental Panel on Climate Change. Cambridge University Press, Cambridge, United Kingdom and New York, USA (2014)

91. Song, W.Y., Mendoza-Cózatl, D.G., Lee, Y., Schroeder, J.I., Ahn, S.N., Lee, H.S., Wicker, T., Martinoia, E.: Phytochelatinmetal (loid) transport into vacuoles shows different substrate preferences in barley and Arabidopsis. Plant Cell Environ. **37**, 1192–1201 (2014). https://doi. org/10.1111/pce.12227

92. Spinelli, F., Fiori, G., Noferini, M., Sprocatti, M., Costa, G.: A novel type of seaweed extract as a natural alternative to the use of iron chelates in strawberry production. Scientiahorticulturae **125**(3), 263–269 (2010). https://doi.org/10.1016/j.scienta.2010.03.011

93. Stirk, W.A., Tarkowská, D., Turečová, V., Strnad, M., Van Staden, J.: Abscisic acid, gibberellins and brassinosteroids in Kelpak®, a commercial seaweed extract made from *Ecklonia maxima*. J. Appl. Phycol. **26**(1), 561–567 (2014). https://doi.org/10.1007/s10811- 013-0062-z

94. Subbarao, P.V., Periyasamy, C., Suresh Kumar, K., Srinivasa Rao, A., Anantharaman, P.: Seaweeds: distribution, production and uses. In: Noor, M.N., Bhatnagar, S.K., Shashi, S.K. (eds.) Bioprospecting of Algae (2018)

95. Sugandhika, M.G.G., Pakeerathan, K., Fernando, W.M.K.: Efficacy of seaweed extract on chilli leaf curl virus. J. Agro-Technol. Rural Sci. **1**. ISSN: 2792-1360 (2021). https://doi.org/ 10.4038/atrsj.v1i1.24

96. Suo, F., You, X., Yin, S., Wu, H., Zhang, C., Yu, X., Sun, R., Li, Y.: Preparation and characterization of biochar derived from co-pyrolysis of *Enteromorpha prolifera* and corn straw and its potential as a soil amendment. Sci. Total Environ. **798**, 149167 (2020). https://doi.org/ 10.1016/j.scitotenv.2021.149167

97. Suresh, S.P., Gauri, S., Desai, S., Kavitha, M., Padmavathy, P.: Marine biotoxins and its detection. Afr. J. Environ. Sci. Technol. **8**(6), 350–365 (2012). https://doi.org/10.5897/AJE ST12.065

98. Thivya, J., Vijayaraghavan, J.: Single and binary sorption of reactive dyes onto red seaweed-derived biochar: multi-component isotherm and modeling. Desalin. Water Treat. **156**, 87–95 (2019). https://doi.org/10.5004/dwt.2019.23974

99. Trivedi, K., Anand, K.V., Vaghela, P., Ghosh, A.: Differential growth, yield and biochemical responses of maize to the exogenous application of *Kappaphycus alvarezii* seaweed extract, at grain-filling stage under normal and drought conditions. Algal Res. **35**, 236–244 (2018). https://doi.org/10.1016/j.algal.2018.08.027

100. Trivedi, K., Anand, K.V., Kubavat, D., Patidar, R., Ghosh, A.: Drought alleviatory potential of *Kappaphycus* seaweed extract and the role of the quaternary ammonium compounds as

its constituents towards imparting drought tolerance in Zea mays L. J. Appl. Phycol. **30**(3), 2001–2015 (2018). https://doi.org/10.1007/s10811-017-1375-0

101. Tsutsui, I., Huybh, Q.N., Nguyen, H.D., Arai, S., Yoshida, T.: The common marine plants of Sothern Vietnam, Japan Seaweed Association, Kochi, 250 pp (2005)

102. Van Ginneken, V., de Vries, E.: Seaweeds as biomonitoring system for heavy metal (HM) accumulation and contamination of our oceans. Am. J. Plant Sci. **9**, 1514–1530 (2018). https://doi.org/10.4236/ajps.2018.97111

103. Veeragurunathan, V., Mantri, V.A., Vizhi, J.M., Eswaran, K.: Influence of commercial farming of *Kappaphycus alvarezii* (Rhodophyta) on native seaweeds of Gulf of Mannar, India: evidence for policy and management recommendation. J. Coast Conserv. **25**, 51 (2021). https://doi.org/10.1007/s11852-021-00836-1

104. Vijayanand, N., Ramya, S.S., Rathinavel, S.: Potential of liquid extracts of Sargassumwightii on growth, biochemical and yield parameters of cluster bean plant. Asian Pac. J. Reprod. **3**(2), 150–155 (2014). https://doi.org/10.1016/S2305-0500(14)60019-1

105. Vijayaraghavan, K., Ashokkumar, T.: Characterization and evaluation of reactive dye adsorption onto biochar derived from *Turbinaria conoides* biomass. Environ. Progress Sustain. Energy, **38**(4) (2019). https://doi.org/10.1002/ep.13143

106. Wan Maznah, W.O.: Perspectives on the use of algae as biological indicators for monitoring and protecting aquatic environments, with special reference to Malaysian freshwater ecosystems. Trop. Life Sci. Res. **21**, 51–67 (2010)

107. Wong, M.H., Chan, K.C., Choy, C.K.: The effect of iron ore tailings on the coastal environment of Tolo Harbour, Hong Kong. Environ. Res. **15**, 342–356 (1978). https://doi.org/10.1016/0013-9351(78)90116-0

108. Wu, H., Huo, Y., Zhang, J., Liu, Y., Zhao, Y., He, P.: Bioremediation efficiency of the largest scale artificial *Porphyra yezoensis* cultivation in the open sea in China. Mar. Pollut. Bull. **95**, 289–296 (2015). https://doi.org/10.1016/j.marpolbul.2015.03.028

109. Xiao, X., Agusti, S., Lin, F., Li, K., Pan, Y., Yu, Y.: Nutrient removal from Chinese coastal waters by large-scale seaweed aquaculture. Sci. Rep. **7**, 46613 (2017). https://doi.org/10.1038/srep46613

110. Xu, C., Leskovar, D.I.: Growth, physiology and yield responses of cabbage to deficit irrigation. Hortic. Sci. **41**(3), 138–146 (2014). https://doi.org/10.17221/208/2013-HORTSCI

111. Xu, C., Leskovar, D.I.: Effects of *A. nodosum* seaweed extracts on spinach growth, physiology and nutrition value under drought stress. Sci. Hortic. **183**, 39–47 (2015). https://doi.org/10.1016/j.scienta.2014.12.004

112. Yang, Y., Chai, Z., Wang, Q., Chen, W., He, Z., Jiang, S.: Cultivation of seaweed *Gracilaria* in Chinese coastal waters and its contribution to environmental improvements. Algal Res. **9**, 236–244 (2015). https://doi.org/10.1016/j.algal.2015.03.017

113. Yang, Y., Liu, Q., Chai, Z., Tang, Y.: Inhibition of marine coastal bloom-forming phytoplankton by commercially cultivated *Gracilaria lemaneiformis* (Rhodophyta). J. Appl. Phycol. **27**, 2341–2352 (2015b). https://doi.org/10.1007/2Fs10811-014-0486-0

114. Yuvaraj, D., Gayathri, P.K.: Impact of seaweeds in agriculture. In: Nabti, E. (ed) Biotechnological Applications of Seaweeds. Nova Science Publishers, Inc., pp. 26–47 (2017)

115. Zheng, Y.M., Liu, T., Jiang, J.W., Yang, L., Fan, Y.P., Wee, A.T.S., Chen, J.P.: Characterization of hexavalent chromium interaction with *Sargassum* by X-rayabsorption fine structure spectroscopy, X-ray photoelectron spectroscopy and quantum chemistry calculation. J. Colloid Interface Sci. **356**(2), 741–748 (2011). https://doi.org/10.1016/j.jcis.2010.12.070

116. Zhou, W., Sui, Z., Wang, J.: Effects of sodium bicarbonate concentration on growth, photosynthesis, and carbonic anhydrase activity of macroalgae *Gracilariopsis lemaneiformis*, *Gracilaria vermiculophylla*, and *Gracilaria chouae* (Gracilariales, Rhodophyta). Photosynth. Res. **128**, 259–270 (2016). https://doi.org/10.1007/s11120-016-0240-3

Chapter 13
Biofertilizer Technologies for Better Crop Nutrient—A Sustainable Smart Agriculture

Aneesha Singh, Bablesh Ranawat, and Monika Rank

Introduction

The global land area of agriculture is about 38% or 5 billion hectares. About 1/3 of this is used as cropland, while the remaining 2/3 consists of pastures and meadows for grazing livestock. China has the biggest agricultural land extent with about 500 million hectares, followed by the United States, Australia, and Brazil. The rising rate of population, especially in Asia, will lead to increased food demand, which in turn will lead to the increased use of fertilizers. Therefore, the regions of Asia and Africa are the largest consumers of fertilizers. The biggest concerns in this region are pollution and contamination of soil.

For the survival of human beings, there are three basic needs food, water, and shelter. Food requires land and fertile land is essential for crop production as our life depends on it. The scarcity of food, increasing demands for food and ceasing nutrient accumulations have consequences in a major deficiency in the food supply and demand chain. It would be awful in the next few years by pressurizing the farmers and agricultural sector to increase the production of food. Agriculture is inseparable from human nourishment and elementary development. In 2015, the UN has acquired 17 objectives with the aim to get rid of hunger and paramount poverty by 2030 while preserving the environment and the global climate. To meet the food grain demand of 7.9 billion people expected to be approximately 10 billion by 2050, grain annual

A. Singh (✉) · B. Ranawat · M. Rank
Academy of Scientific and Innovative Research (AcSIR), CSIR-HRDC Campus, Sector 19, Kamla Nehru Nagar, Ghaziabad, Uttar Pradesh 201002, India
e-mail: aneeshas@csmcri.res.in

M. Rank
e-mail: monikarank@csmcri.org

Applied Phycology and Biotechnology Division, CSIR-Central Salt and Marine Chemicals Research Institute, Bhavnagar, Gujarat 364002, India

© Centre for Science and Technology of the Non-aligned and Other Developing Countries 2023
K. Pakeerathan (ed.), *Smart Agriculture for Developing Nations*, Advanced Technologies and Societal Change, https://doi.org/10.1007/978-981-19-8738-0_13

production raise of 50% is needed [1]. This implicates the sustainable enhancement of persisted agricultural land by innovation and multi-sectoral collaboration. Therefore, it is important to use available agricultural land sustainably to fulfil the food need of the large population of the world [2]. Farming is an earning source for people in India and many other developing countries.

In India, desertification, deforestation, soil erosion, drought, water logging, flooding, and over-exploitation of soils is due to a rise in standards of living comfort and invasion of agricultural land due to urbanization and transportation facility. Such soils are highly deficient in plant nutrients, humus, and other organic materials. The fertility of Indian soils has been lost because they have been used for cultivation for decades. There is an occurrence of 8 different types of soils available in different regions in India, such as black soils, laterite soils, alluvial soils, peaty soils, saline and alkali soil and red soil, arid and desert soil, and forest and mountain soil (Fig. 13.1; and edge.com). In alkaline and saline soils, the topmost upper fertile layer is drenched with salt particles. Weathering of rock gives rise to sodium, magnesium, calcium salts, and sulphurous acid. Saline soils have weak structural stability, low hydraulic conductivity, and infiltration rate that results in decreased crop productivity. In India, this type of soil occurs in Gujarat, Andhra Pradesh, Bihar, Telangana, Haryana, Karnataka, Punjab, Uttar Pradesh, Rajasthan, and Maharashtra. Along the coastline, the soil is unfit for cultivation due to storm surges and saline seawaters infiltrating coastal regions. The application of gypsum is the only way to treat alkaline soil to reduce pH. The repeated use of gypsum may not be beneficial in long term. Limestone and mudstones are common phosphate-bearing rocks, and bacteria can solubilize insoluble minerals in the soil [4] and make them available. Healthy crop production required fertile soil rich in macronutrients and micronutrients for good crop production. Plants absorb these nutrients and minerals from the soil by roots with the course of time the agricultural land becomes deficient in these nutrients and minerals. Farmers add these nutrients to the soil externally through chemical fertilizers. Chemical fertilizers act as a double-edged sword, it provides nutrients to plants quickly, but their prolonged use adversely affects soil fertility. In addition, their absorption is relatively low [5], and its overuse leads to the depletion of essential nutrients with deleterious effects on the water quality and atmospheric gases [6]. The application of fertilizer is necessary to increase food production and maintain soil fertility for continuous farming.

Microbes help in plant growth promotion by raising the availability of essential nutrients and stress tolerance [7–9]. Biofertilizers application minimizes the additional application of chemical fertilizer and soil softener [10], and therefore, the cultivation cost is reduced. To utilize the unavailable minerals already present in the soil and combat the deleterious effects of synthetic chemical fertilizers, there is a need to use eco-friendly methods like biofertilizers and organic manure. These methods are ecologically acceptable as well as economically sustainable for agriculture [11].

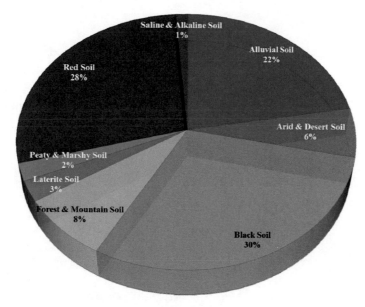

Fig. 13.1   Occurrence of different types of soils in India

Biofertilizer

Biofertilizers are microbial inoculums containing a variety of living cells or dormant states of efficient microorganisms that promote plant growth, nutrient uptake efficiency, and soil fertility. Microbes have multiple functions (Fig. 13.2) adding nutrients, replenishment, and soil enrichment, softening, and nourishing [12]. There is a plethora of studies that have reported the significance of algae from diverse sources as a potential biofertilizer. Algal biofertilizers are a good nutrient source and enhance plant growth [13]. Biofertilizers introduce nutrients through the natural and spontaneous processes of N_2 fixation, and demineralization of natural minerals rich in phosphorous, potassium, calcite, iron, etc., and enhancing crop yield by the breakdown of growth-enhancing compounds.

Isolation, Screening, and Identification of Microbiome

Samples are collected from appropriate sites and cultured on nutrient agar medium. After incubation, morphologically different colonies are isolated on the same medium. These isolates are further tested for phosphate solubilization on Pikovskaya's medium, potassium solubilization on Aleksandrow medium, and calcite solubilization on calcite solubilizing medium. A zone of clearance represents solubilization by the isolate. The capability of nitrogen fixation can be detected

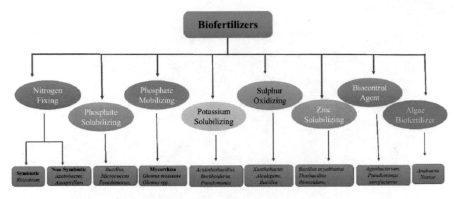

Fig. 13.2 Various known biofertilizers used nowadays

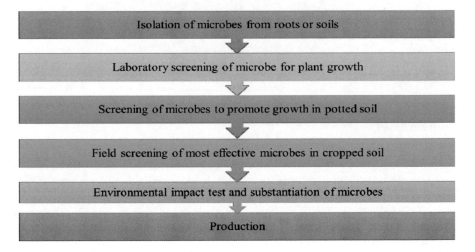

Fig. 13.3 Flowchart depicted the steps involved in the development of a biofertilizer

on Burk's modified *N*-free medium. Biochemical characteristics of bacterial isolates can be studied by using indole test, methyl red test, Voges–Proskauer (VP) test, and citrate utilization test. The isolates identified by the 16S rRNA gene using NCBI BLAST [7–9, 14]. After getting desired and effective strain, these can be further tested in potting soil for promoting plant growth (Fig. 13.3).

Selection of Artificial Microbial Community

There are two techniques for the selection of microbial stains; the first one is to modify existing microbial communities, and the second is to build synthetic communities (SynComs). In the first one, abiotic factors are used to manipulate native microbes as per requirement. In the second case, pre-selected microbes are used to build up an artificial community [15]. The most challenging part is to identify keystone taxa that are highly influential on the native community. Based on metagenomics and sequencing data, an overview of their interaction can be understood. Artificial microbial consortia (AMC) have the advantage of the addition and deletion of microbes under controlled conditions. Less complex-Syncoms has the disadvantage of missing some important association that is critical for field application, and complex-Syncoms had some design limitation, but the microbial association is better and intact [16].

Plant Probiotics and Customized Inoculant

Application of favourable microbes for plant growth and development acts as probiotics. Signalling molecules can be added to soil to promote soil microbes. They can be combined with biofertilizer to further enhance their efficiency and use of root exudates to attract plant associate microbes. Customized farming practice is based on fertilizer consultants for long-lasting effects [17]. Production of mycorrhizae inoculum on the farm is tested on potatoes [18]. However, the signal formula for all fields seems to be impractical. Such strategies are low cost, but their global feasibility seems to be difficult.

Biofilm Biofertilizers

Microbes that are capable to produce biofilm are used in biofilm biofertilizers. Extracellular polymeric substance exudates by them provide protection and synergies with other microbial communities and plants. Biofilm provides a suitable environment that helps in combating with native microbial community. Microbial consortia biofilm is better than single microbe biofilm inoculums [19]. Bacterial–fungal biofilm was found to be better in abiotic stress tolerance as compared to single or consortium of non-biofilm producing microbes [20]. These microbes strongly adhere to biotic and abiotic molecules and are thereby capable of enhancing fertility in soil and stress tolerance in plants.

Methods of Biofertilizer Applications

1. Inoculant is blended with the solution of jaggery and rice gruel. The seeds are mixed well with the slurry for even application of inoculants (*Azotobacter*, *Azospirillum*, and *Rhizobium*) over the seeds. After drying for half an hour in shade, they are used for sowing, within 24 h.

2. The part of the seedlings root is dipped in the solution for 5–10 min containing the desired inoculant, and the seedlings are transplanted into the field.
3. Desired inoculants mixed with compost, kept overnight, and applied at the time of sowing or planting (Fig. 13.4). Various biofertilizer application methods are applied on different crop varieties (Table 13.1).

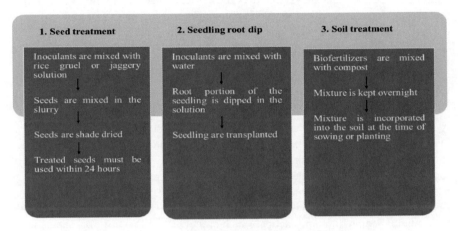

Fig. 13.4 Schematic diagram represents the treatment of the biofertilizer

Table 13.1 An overview of biofertilizer application methods on various crops

S. No.	Application	Applicable to crop varieties	Benefits	References
1	Seed treatment	Cereals, pulses	Economic, effective, general method for all types of inoculants	Taylor and Harman [21] Deaker et al. [22] Bashan et al. [23] Sethi et al. [24]
2	Seedling root dip	Cereals, fruits, vegetables, trees, cotton, banana, grapes, sugarcane, tobacco	Used for plantation crops	Barea and Brown [25] Youssef and Eissa [26] Garg et al. [27]
3	Soil treatment	Fruits, tea, coffee, coconut, rubber, spice, nuts, flowers	Applicable either single or in combination directly to the soil	Zahran [28] Hayat et al. [29] Bashan et al. [23]

Case Study

Application of *Enterobacter Hormaechei* on Tomato Plants

Effect of *Enterobacter hormaechei* Application on Root Growth and Architecture

Root growth and architecture changed by the application of *Enterobacter hormaechei* (Fig. 13.5a–k). The tomato seedlings were cultured in hydroponic system with and without 25–50–75 mg/100 ml K-feldspar, 25–50–75 mg/100 ml calcium carbonate, and 25–50–75 mg/100 ml tri-calcium phosphate and *E. hormaechei*. The isolate was able to solubilize calcite, potassium, phosphate, fix free nitrogen, and IAA production [7–9]. Early germination with better growth vigour was observed in *E. Hormaechei-*treated seedlings (Fig. 13.5a). All the three calcium carbonate-treated seedlings have better root growth as compared to control (Fig. 13.5b–d). Significant differences were observed in shoot, leaves, and root in calcium carbonate-treated plants as compared to control (Fig. 13.6a). Tri-calcium phosphate-treated seedlings have better shoot and root growth as compared to control (Fig. 13.5e–g). However, there was no significant difference in the root length in treated seedling as compared to control (Fig. 13.6b). K-feldspar-treated seedlings have better shoot, leaves, and root growth as compared to control seedlings (Figs. 13.5h and 13.6c). The synergistic effect of K-feldspar and tri-calcium phosphate increased both, number of roots and root length, along with the shoot length (Fig. 13.5i–k). These results reveals that co-cultivation of *E. hormaechei* improved root growth and architecture. The results were in corroborate with Ranawat et al. [7–9].

Enhanced Crop Production Under Salinity Stress and Disease Tolerance in Tomato Plants by *E. hormaechei*

The pot trial experiment of tomatoes was carried in the greenhouse (Fig. 13.7). The plant's pot soil was treated with the isolate and different concentrations of salinity. Better growth and biomass were obtained in those plants treated with salinity + *E. hormaechei* + K-feldspar in contrast to those positive (Fig. 13.7A) and negative (Fig. 13.7B) control. The fruit ripening started first in K-field spar-treated plants followed by tri-calcium phosphate-treated plants (Fig. 13.7C–D). The poor plant growth and yield and diseased tomatoes were obtained from control plants treated with sanity without isolate [7]. The application of *E. hormaechei* helps to solubilize and convert this unavailable mineral complex form to simpler available form of nutrients that are up taken by plants. Therefore, microbes play a vital role in increasing in plant growth and productivity. In this case study, *E. hormaechei* was capable of solubilize phosphate, potassium, and calcium, fix free nitrogen, produce IAA, and therefore, crop yield significantly increased by the application of *E. hormaechei* [7–9].

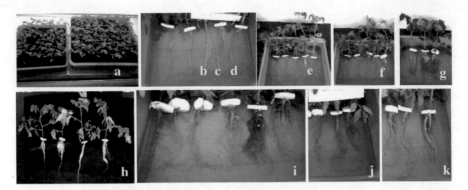

Fig. 13.5 Co-cultivation of *Enterobacter hormaechei* on tomato seedling; left side tray of control seedlings and right side tray of *E. hormaechei*-treated seedlings (**a**), all of the 3 seedlings in the left side tray are of control and right side are seedlings treated with *E. hormaechei* + calcium carbonate, 25 mg/100 ml (**b**), 50 mg/100 ml calcium carbonate (**c**), and 75 mg/100 ml calcium carbonate (**d**), seedlings treated with *E. hormaechei* + 25–50–75 mg/100 ml tri-calcium phosphate (**e–g**), seedling treated with *E. hormaechei* + 25–50–75 mg/100 ml feldspar in right side, and control in the left side of the tray (**h**), seedling treated with *E. hormaechei* containing 25 mg/100 ml feldspar + 25 mg/100 ml tri-calcium phosphate levels on right hand side and control on the left side of the tray (**i**), synergistic effect of *E. hormaechei* + 50 mg/100 ml feldspar + 50 mg/100 ml tri-calcium phosphate on tomato seedling (**j**), synergistic effect of *E. hormaechei* + 75 mg/100 ml feldspar + 75 mg/100 ml tri-calcium phosphate on tomato seedling (**k**)

Biofertilizer Formulation and Applications: New Technologies

Biofertilizer formulation development strategies have changed in the recent past. The selection strain of microorganisms that have more than one solubilizing traits, and the use of microbial consortia instead of a single organism is under practice [7–9]. Microbial consortia consist of closely or distantly related two or more strains that anticipate an overall supplementing or collaborative effect of biofertilization [30, 31]. In addition to it, the combined use of rhizobacteria and arbuscular mycorrhizal fungi (AMF) is in practice. The use of isolation chips [32] and the diffusion chambers [33, 34] that imitate natural conditions raised the counting of cultured colonies and marked a revival of culture-dependent methods. The focus of the culturomics method is to suppress the inoculum of fast-growing bacteria to further enhance their growth [35]. Lagier et al. [35, 36] had studied "Culturomics," there is a variety of combinations in culturomics that atmospheres, culturing conditions, times of incubation and various growth media have to be developed for the plant and soil-associated microbiota. A "plant-tailored culturomics method" that combines culturomics with plant-based media using animal nutrients [37]. Most of studies regularly practise using routine general media that contains animal originated nutrients, for example Luria–Bertani, nutrient agar, and R2A to isolate plant-associated microbes, wherever dehydrated juices powders or plant materials should be used alternatively. For the suitable formulation, there is need of microbe of choice and their functions, to assure microbial cell viability during application and storage. There are some limitations in

Fig. 13.6 Effect of *E. hormaechei* co-cultivation on tomato seedlings grown in hydroponics; **a** seedlings were treated with different concentrations of calcium carbonate + *E. hormaechei*, **b** seedlings treated with tri-calcium phosphate + *E. hormaechei*, **c** seedling treated with K-feldspar + *E. hormaechei* at various concentration compared with their respective control; the statistical analysis by two-way ANOVA

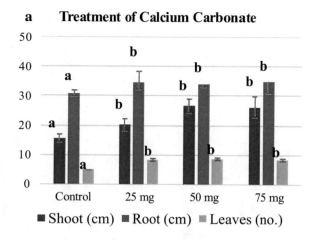

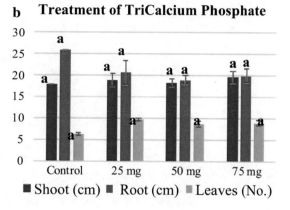

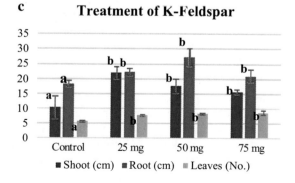

Fig. 13.7 Stress tolerance and crop enhancement in tomatoes by the co-cultivation of *E. hormaechei* under greenhouse conditions; **a** treated with salinity, **b** treated with *E. hormaechei* + salinity, **c** treated with *E. hormaechei* + salinity + feldspar, **d** treated with *E. hormaechei* + salinity + tri-calcium phosphate

both the solid carrier-based formulation and liquid formulation, so it can be lead to greater setbacks for the commercial market. To overcome this problem, two different strategies are available such as drying processes using a fluid bed dryer and microbial encapsulation with polymeric hydrogels. These polymeric hydrogels supply firstly a diffusion obstacle, which permits the pathway to molecules with optimum size and an aqueous environment, which assists to support the activity of the microbes [38]. There are many natural (polysaccharides) and synthetic (polyurethane, poly-acrylamide) polymers that have been extensively used for microbial encapsulation. Alginates are largely used due to their less toxicity, ease of gelation, and excessive biocompatibility in assisting the survival of cell [39]. There are some limitations in the use of these carriers in crop cultivation, such as naturally available polymers have low mechanical strength and heat sensitive in contrast to artificial polymers [40]. Another drawback is the porosity of the alginate restricts industrial application. Moreover, alginate is relatively expensive [41, 42]. A fluidized bed dryer (FBD) is a novel technique for formulation development to increase the survival rate of an inoculant and reduce the risk of contamination. This method has been enormously used in food and pharmaceuticals to decrease the water content of granules and powders [43]. The major benefit of this method is that it operates ~ 37 to 40 °C temperature, which is highly fit for certain groups of organisms [44]. There are several studies using FBD that have shown plant growth promotion [45] and disease tolerance [46–48]. This FBD technology is not commonly used in microbial inoculants formulations. There is a need to optimize the temperature cycles of FBD for the formulation and to assess its rate of survival after application in field. So far, both processes using alginate and FBD have a good effect in decreasing inconsistency of formulation by maintaining cell density of microbes in course of storage time period. Moreover, these methods may unlock new possible criteria to increase the storage capacity of non-spore-forming bacteria.

Nanobiofertilizer

The formulation of nanofertilizer facilitates the better nutrient absorption and retards nutritional deprivation. Rajonee et al. [49] reported formulation and synthesis of nanofertilizers using ammonium salts, phosphorus, potassium, and urea. The simple and eco-friendly nanonutrients formulation derived from plant biomass and microbes are also reported [50]. Various nutrient particles are implanted with nanofunctionalized materials by a thin coating of nutrients on the functionalized materials surface, and encapsulation of nutrients within the nanofuctionalized nanoscale suspensions distributed in emulsions. Such nano-coats evade unwanted interactions of the nutrients with microorganisms and provide surface protection. There are many reports that have indicated nanomaterial's advantageous role in regulating unnecessary and undesirable infections and contamination [51]. Silica nanomaterial reported to be advantageous in the nourishment of seed and development of root. Some preliminary studies have reported the enhancement in the growth and development of crops under in vitro conditions. Blending of nanoparticles with *Pseudomonas* spp. and *Bacillus* spp. has given beneficial effects [52]. The treatment of zinc oxide nanoparticles on peanuts was advantageous in the sprouting of seeds, early flowering, faster shoot, and root growth, more chlorophyll content in leaf, with higher pod yield [53]. Salama [54], reported similar results when treatment of silver nanoparticles on wheat, intimated better shoot and root elongation, protein, starch, and pigment content. There are many biofertilizers and nanobiofertilizer available in the market as long-lasting, cost-effective, and promising alternative options to chemical fertilizers for sustainable agriculture. Soil-less agriculture techniques are emerging like hydroponics and aeroponics using terrace garden farming.

Biofertilizer and Biosafety

There is a need to understand the risks to human health before the introduction of plant growth promoting microorganisms in large-scale agriculture.

Risk on Health of Human Being

In the past, a variety of PGP bacteria live in association with roots; out of them, some are reported to be pathogens for plants as well as for human beings [55]. There are many human pathogens belonging to various genera including *Acinetobacter, Bacillus, Pseudomonas, Enterobacter, Klebsiella, Burkholderia, Serratia, Ochrobactrum, Rhodococcus, Ralstonia, Stenotrophomonas* reported having high PGP activity [56, 57]. In addition, [58] have studied the bacterial genus *Pseudomonas*, which includes various species of PGP such as *P. putida, P. fluroescens, P. putrefaciens, P. pseudoalcaligens, and P. stutzeri*, however, *P. aeruginosa* an opportunistic pathogenic species that cause infections of the respiratory tract in humans. Recently, the highly available biofertilizers in the market consist of the formulation of nitrogen-fixing organisms that have less risk of health and are known to be safe for field applications [57, 59]. According to the above-mentioned information, it is clear that biosafety screening is required for any PGPB strain and assures the product safety, prior to taking to commercial scale [55, 56]. Non-pathogen PGPB can be used that is safe for the health of humans and the environment.

Effects on Environmental Health Risks

The impact of the introduction of new microorganisms on the native microbes is of great concern [60, 61]. The introduction of microorganisms can indirectly modify native microbial populations, by modifying plant morphology and physiology. In combined plant populations, it indirectly affects the native microbiota [56, 62]. The effect of new microbiomes on the native microbiota is subjected to their capacity to live in the environment [63]. Plant genotype alteration by the application of microbes has been reported [63, 64]. *A. brasilense* co-cultivation controlled the phylogenetic, phyllospheric, and rhizospheric changes. Another study [65] reported that the response of the rhizoplane microbiome to affected at the grain-filling stage. On the other hand, the existing native microbial communities might have different levels of buffering capacity and susceptibility depends on their diversity [66, 67] and the occurrence of friendly and non-friendly groups of microbes [68, 69]. Climatic conditions may also play a vital step in maintaining the effect of implied microbes [64, 70, 71]. Further study and research is required to completely reveal the effect of microbial consortia on the resident microbiome.

Current Technology Bottleneck

Biofertilizers have been the centre of the research discussion for a decade. There has been a step-by-step change from chemical fertilizer to several biofertilization techniques [72]. The source of organic bio-based fertilizers can be plants, animals, and microorganisms [74]. The main component of these bio-based organic fertilizers is live microbial biomass. The major concern in the agricultural inoculation techniques is the survival of microorganisms on storage. Various parameters affect biofertilizer efficiency, for example culture medium, microbial physiology, dehydration process, drying rate, maintenance of temperature during storage, and water activity of inoculants. The powder carrier-based formulation biofertilizers shelf life is generally limited to 6–12 months, and liquid-based biofertilizers shelf life is only 3–6 months [74]. The development of biofertilizer formulation was described as from small-scale studies in the laboratory by in vitro practices to successful practice leads to pilot scale, after that on large-scale plant production that led to field application (Fig. 13.3). Only, 0.1% of PGPR species has been developed for commercial use so far indicating rare attempts in the area of microbial biofertilizer formulations. Also, there is lack of awareness and knowledge in the farmers with less adoption rate to biofertilizers.

Marketing of Biofertilizers: Global Scenario

The market for biofertilizers was estimated to be worth US $2.6 billion globally in 2021, and it is anticipated to reach US $4.5 billion by 2026, rising at a compound annual growth rate (CAGR) of 11.9%. Currently, consumers are getting largely concerned and aware of food safety issues, environmental issues, and the increasing residue levels in food because of the increasing concerns about their fitness and well-being. This increase in consciousness and concern has prompted them to favour organic food products. As a consequence, chains of prime supermarkets such as Walmart and Cosco are enlarging their products by providing chemical-free foods. Since organic manures and biofertilizers are essential to organic farming, the demand for them is growing along with the organic food business. The COVID-19 pandemic's emergence has increased consumer interest in natural, chemical-free, nutritious food items, which has fuelled the demand for biological insertions like biofertilizers. According to type, the nitrogen-fixing microbes market is expected to have the greatest CAGR throughout the forecast period. In addition to the necessary microbes and their nutrients, liquid biofertilizers also contain ingredients that can help maintain the conditions of storage of the dormant phase for a longer preservation time. Liquid formulations are an optional choice to the traditional form of carrier-based biofertilizers and offer higher tolerance limits for unfavourable situations. With these advantages, liquid biofertilizers were predicted to hold the biggest market share in terms of form. Because seed treatment is the most common technique of administering biofertilizers, it is simple and typically efficient method in most situations. It

Table 13.2 List of key market players of the world

S. No.	Market players	Country
1	AgriLife	India
2	Agrinos	US
3	Aumgene Biosciences	India
4	Biomax Naturals	India
5	Chr. Hansen Holding A/S	Denmark
6	Criyagen	India
7	IPL Biologicals Limited	India
8	Jaipur Bio Fertilizers	India
9	KanBiosys	India
10	Kiwa Bio-Tech	China
11	Lallemand Inc.	US
12	LKB BioFertilizer	Malaysia
13	Manidharma Biotech Pvt Ltd.	India
14	Mapleton Agri Biotech Pty Ltd	Australia
15	Novozymes A/S	Denmark
16	Nutramax Laboratories Inc.	US
17	Rizobacter Argentina S.A	Argentina
18	Seipasa	Spain
19	Symborg	Spain
20	T. Stanes & Company Limited	India
21	UPL Limited	India
22	Valagro	Italy
23	ValentBioSciences	US
24	Varsha Bioscience and Technology India Pvt Ltd.	India
25	Vegalab SA	Switzerland

is predicted that this segment would see the highest CAGR over the course of the projection period (Table 13.2).

Government Schemes Promoting Biofertilizers: Prospects in India

The Indian government promotes the production and use of biofertilizers. Goverment is spreading awareness among farmers by arranging camps seminars, and training programs. The policy of marketing biofertilizers at very low prices attracts sufficient investors and manufacturers. The Indian government has introduced various schemes supported by NABARD, which are encouraging the use of biofertilizers (www.agrifarming.in).

1. Paramparagat Krishi Vikas Yojana (PKVY)
2. Mission Organic Value Chain Development for North Eastern region (MOVCDNER)
3. National Scheme Mission on Oilseeds and oil Palm (NMOOP)
4. National Food Security Mission (NFSM)
5. National Project on Organic Framing (NPOF)
6. National Horticultural Mission (NHM)
7. Rashtriya Krishi Vikas Yojana (RKVY).

Terrace Gardening

India is an agricultural country; 70% of the population here is directly or indirectly dependent on agriculture. But, most of the agriculture practices are by using chemical fertilizers. Hence, people are searching for new methods to get their own vegetables for daily use. In metro cities, due to a lack of land, people are diverting towards hydroponic vertical farming to grow their own vegetables in the terrace garden. Hydroponics can be undertaken as greenhouse farming, rooftop farming, and indoor farming. High productivity can be obtained by the optimization of pH, light, temperature, irrigation, humidity, and CO_2, [75]. Compared to conventional agriculture new automated Controlled Environment Agriculture (CEA) hydroponic has high crop yield in small spaces. However, the application of biofertilizers, cost cutting, and waste management need more research to popularize this method.

Aeroponics

It is an advanced greenhouse approach, one step ahead of hydroponic. In this approach, an enclosed air system comprised of excessive humidity in which plant roots are suspended in air, and a fine spray of nutrient solution is provided like aerosol or mist [42]. The enclosed system requires a water-mix solution as a feed to plant roots. The water recycling in the system results in 95% water saving compared to conventional farming. Arbuscular mycorrhiza has more potential for use as a biofertilizer in aeroponics systems. The aeroponic culture of AMF permits both efficient productions of Arbuscular mycorrhiza fungus inoculum and soil-free investigation of mycorrhiza. The production of AMF in a culture of aeroponic permits easy extractions of roots [76]. In this system, plants grow healthier and faster, due to the better environmental conditions. But, in most cases, environments of aeroponic are not perfectly closed off to the outside, hence, pests and diseases may still cause a threat.

Conclusion

The use of biofertilizers and nanofertilizer with modern techniques like hydroponics, aeroponic, and terrace garden farming for sustainable agriculture enabled to encounter the demand for increased food that prevents starvation and malnutrition. It is well understood by hydroponic studies; *E. hormaechei* enhanced crop yield due to its potential capacity to solubilize minerals. This might be possible due to changes in root architecture that have increased nutrient uptake. The effective use of PGPBs as potential plant nutrients remains a dream until these benefits are not percolated to end users. In addition, limitations in biofertilizer technology like shelf life, technology gaps, high risk of contamination, and the non-viability of microorganisms when exposed to high-temperature need to be solved. Therefore, there is a need to search for beneficial bacterial isolates and develop microbial consortia to check their stability in the environment to promote and boost crop production. Also, develop new techniques to lower the pressure on agricultural land. Moreover, due to the prohibition on deleterious chemical pesticides, the growth of the market for biofertilizers will not drop but may increase by time. For sustainable agriculture, there is a need to focus research on the potential bacterial strains, with high shelf life, and enhancing the efficiency of existing isolates using biotechnological methods.

Acknowledgements CSIR-CSMCRI Communication No 174/2022. The no financial support to declare.

References

1. Mahapatra, D.M., Satapathy, K.C., Panda, B.: Biofertilizers and nanofertilizers for sustainable agriculture: phycoprospects and challenges. Sci. Total Environ. **803**, 149990 (2022)
2. Chabbi, A., Lehmann, J., Ciais, P., Loescher, H.W., Cotrufo, M.F., Don, A., SanClements, M., Schipper, L., Six, J., Smith, P., Rumpel, C.: Aligning agriculture and climate policy. Nat. Clim. Chang. **7**, 307–309 (2017)
3. https://andedge.com/types-of-soil-in-india (20 July 2022)
4. Cacchio, P., Ferrini, G., Ercole, C., Del Gallo, M., Lepidi, A.: Biogenicity and characterization of moonmilk in the grottanera (Majella National Park, Abruzzi, Central Italy). J. Cave Karst Stud. **76**, 88–103 (2014)
5. Trenkel, M.E.: Improving Fertilizer Use Efficiency: Controlled-Release and Stabilized Fertilizers in Agriculture, The International Fertilizer Industry Association, p. 151. International Fertilizer Industry Association (IFA), Paris, France, Paris (1997)
6. Haygarth, P.M., Bardgett, R.D., Condron, L.M.: Phosphorus and nitrogen cycles and their management. In: Gregory, P.J., Nortcliff, S. (eds.) Soil Conditions and Plant Growth, pp. 132–159. Wiley-Blackwell, West Sussex, UK (2013)
7. Ranawat, B.: Studies on halo-tolerant bacteria for agriculture application. Acsir Ph.D. thesis (2021a)
8. Ranawat, B., Bachani, P., Singh, A., Mishra, S.: *Enterobacter hormaechei* as Plant Growth-Promoting Bacteria for Improvement in *Lycopersicumesculentum*. Curr. Microbiol. **78**, 1208–1217 (2021)

9. Ranawat, B., Mishra, S., Singh, A.: *Enterobacter hormaechei* (MF957335) enhanced yield, disease and salinity tolerance in tomato. Arch. Microbiol. **13**, 1–9 (2021)
10. Thomas, L., Singh, I.: Microbial biofertilizers: types and applications. In: Biofertilizers for Sustainable Agriculture and Environment, pp. 1–19. Springer, Cham (2019)
11. Nosheen, S., Ajmal, I., Song, Y.: Microbes as biofertilizers, a potential approach for sustainable crop production. Sustainability **13**, 1868 (2021)
12. Bhardwaj, D., Ansari, M.W., Sahoo, R.K., Tuteja, N.: Biofertilizers function as key player in sustainable agriculture by improving soil fertility, plant tolerance and crop productivity. Microb. Cell Fact. **13**, 1–10 (2014)
13. Mahapatra, D.M., Chanakya, H.N., Joshi, N.V., Ramachandra, T.V., Murthy, G.S.: Algae-based biofertilizers: a biorefinery approach. In: Panpatte, D., Jhala, Y., Shelat, H., Vyas, R. (eds.) Microorganisms for Green Revolution, vol. 7, pp. 177–196. Springer, Singapore (2018)
14. Mitter, E.K., Tosi, M., Obregón, D., Dunfield, K.E., Germida, J.J.: Rethinking crop nutrition in times of modern microbiology: innovative biofertilizer technologies. Front. Sustain. Food Syst. **5**, 606815 (2021)
15. Raaijmakers, J.M.: The minimal rhizosphere microbiome. In: Lugtenberg, B. (ed.) Principles of Plant-Microbe Interactions, pp. 411–417. Springer International Publishing, Cham (2015)
16. Vorholt, J.A., Vogel, C., Carlstrom, C.I., Muller, D.B.: Establishing causality: opportunities of synthetic communities for plant microbiome research. Cell Host Microb. **22**, 142–155 (2017)
17. Bell, T.H., Kaminsky, L.M., Gugino, B.K., Carlson, J.E., Malik, R.J., Hockett, K.L., et al.: Factoring ecological, societal, and economic considerations into inoculant development. Trends Biotechnol. **37**, 572–573 (2019)
18. Goetten, L.C., Moretto, G., Sturmer, S.L.: Influence of arbuscular mycorrhizal fungi inoculum produced on-farm and phosphorus on growth and nutrition of native woody plant species from Brazil. Acta BotanicaBrasilica **30**, 9–16 (2016)
19. Velmourougane, K., Prasanna, R., Saxena, A.K.: Agriculturally important microbial biofilms: present status and future prospects. J. Basic Microbiol. **57**, 548–573 (2017)
20. Hassani, M.A., Duran, P., Hacquard, S.: Microbial interactions within the plant holobiont. Microbiome **6**, 58 (2018)
21. Taylor, A.G., Harman, G.E.: Concepts and technologies of selected seed treatments. Annu. Rev. Phytopathol. **28**, 321–339 (1990)
22. Deaker, R., Roughley, R.J., Kennedy, I.R.: Legume seed inoculation technology—a review. Soil Biol. Biochem. **36**, 1275–1288 (2004)
23. Bashan, Y., de-Bashan, L.E., Prabhu, S.R., Hernandez, J.P.: Advances in plant growth-promoting bacterial inoculant technology: formulations and practical perspectives (1998–2013). Plant Soil **378**, 1–33 (2014)
24. Sethi, S.K., Sahu, J.K., Adhikary, S.P.: Microbial biofertilizers and their pilot-scale production. In: Microb. Biotechnol. 312–331 (2018)
25. Barea, J.M., Brown, M.E.: Effects on plant growth produced by *Azotobacter paspali* related to synthesis of plant growth regulating substances. J. Appl. Bacteriol. **37**(4), 583–593 (1974)
26. Youssef, M.M.A., Eissa, M.F.M.: Biofertilizers and their role in management of plant parasitic nematodes. A review. J. Biotechnol. Pharm. Res. **5**, 1–6 (2014)
27. Garg, N., Pandove, G., Kalia, A., Pandey, V., Mahala, P.: Differential effects of plant growth-promoting rhizobacteria used as soil application vis-a-vis root dip of seedlings on the performance of onion (*Allium cepa*) in three distinct agro-climatic zones of Indian Punjab. Commun. Soil Sci. Plant Anal. 1–20 (2022)
28. Zahran, H.H.: Rhizobium-legume symbiosis and nitrogen fixation under severe conditions and in an arid climate. Microbiol. Mol. Biol. Rev. **63**, 968–989 (1999)
29. Hayat, R., Ali, S., Amara, U., Khalid, R., Ahmed, I.: Soil beneficial bacteria and their role in plant growth promotion: a review. Ann. Microbial. **60**(4), 579–598 (2010)
30. El Maaloum, S., Elabed, A., Alaoui-Talibi, Z., El, Meddich, A., Filali-Maltouf, A., Douira, A., et al.: Effect of arbuscular mycorrhizal fungi and phosphate-solubilizing bacteria consortia associated with phosphocompost on phosphorus solubilization and growth of tomato seedlings (*Solanumlycopersicum*). Commun. Soil Sci. Plant Anal. **51**, 622–634 (2020)

31. Ramirez-Lopez, C., Esparza-Garcia, F.J., Ferrera-Cerrato, R., Alarcon, A., Canizares-Villanueva, R.O.: Short-term effects of a photosynthetic microbial consortium and nitrogen fertilization on soil chemical properties, growth, and yield of wheat under greenhouse conditions. J. Appl. Phycol. **31**, 3617–3624 (2019)

32. Nichols, D., Cahoon, N., Trakhtenberg, E.M., Pham, L., Mehta, A., Belanger, A., et al.: Use of ichip for high-throughput in situ cultivation of uncultivable microbial species. Appl. Environ. Microbiol. **76**, 2445–2450 (2010)

33. Bollmann, A., Lewis, K., Epstein, S.S.: Incubation of environmental samples in a diffusion chamber increases the diversity of recovered isolates. Appl. Environ. Microbiol. **73**, 6386–6390 (2007)

34. Kaeberlein, T.: Isolating uncultivable microorganisms in pure culture in a simulated natural environment. Science **296**, 1127–1129 (2002)

35. Lagier, J.C., Hugon, P., Khelaifia, S., Fournier, P.E., La Scola, B., Raoult, D.: The rebirth of culture in microbiology through the example of culturomics to study human gut microbiota. Clin. Microbiol. Rev. **28**, 237–264 (2015)

36. Lagier, J.C., Khelaifia, S., Alou, M.T., Ndongo, S., Dione, N., Hugon, P., et al.: Culture of previously uncultured members of the human gut microbiota by culturomics. Nat. Microbiol. **1**, 16203 (2016)

37. Sarhan, M.S., Hamza, M.A., Youssef, H.H., Patz, S., Becker, M., ElSawey, H., et al.: Culturomics of the plant prokaryotic microbiome and the dawn of plant-based culture media—a review. J. Adv. Res. **19**, 15–27 (2019)

38. Perez-Luna, V., Gonzalez-Reynoso, O.: Encapsulation of biological agents in hydrogels for therapeutic applications. Gels **4**, 61 (2018)

39. Gasperini, L., Mano, J.F., Reis, R.L.: Natural polymers for the microencapsulation of cells. J. R. Soc. Interface **11**, 20140817 (2014)

40. Zhu, Y.: Immobilized cell fermentation for production of chemicals and fuels. In: Yang, S.T. (ed.) Bioprocessing for Value-Added Products from Renewable Resources, pp. 373–396. Elsevier, Amsterdam (2007)

41. Reis, C.P., Neufeld, R.J., Vilela, S., Ribeiro, A.J., Veiga, F.: Review and current status of emulsion/dispersion technology using an internal gelation process for the design of alginate particles. J. Microencapsul. **23**, 245–257 (2006)

42. Sahu, K., Mazumdar, S.: Digitally greenhouse monitoring and controlling of system based on embedded system. Int. J. Sci. Eng. Res. **3**, 2229–5518 (2012)

43. Dewettinck, K., Huyghebaert, A.: Fluidized bed coating in food technology. Trends Food Sci. Technol. **10**, 163–168 (1999)

44. Sahu, P.K., Gupta, A., Singh, M., Mehrotra, P., Brahmaprakash, G.P.: Bioformulation and fluid bed drying: a new approach towards an improved biofertilizer formulation. In: Sengar, R.S., Singh, A. (eds.) Eco-Friendly Agro-Biological Techniques for Enhancing Crop Productivity, pp. 47–62. Springer, Singapore (2018)

45. Gangaraddi, V., Brahmaprakash, G.: Evaluation of selected microbial consortium formulations on growth of green gram (*Vignaradiata*). Int. J. Chem. Stud. **6**, 1909–1913 (2018)

46. Gotor-Vila, A., Usall, J., Torres, R., Abadias, M., Teixido, N.: Formulation of the biocontrol agent Bacillus amyloliquefaciens CPA-8 using different approaches: liquid, freeze-drying and fluid-bed spray-drying. Biocontrol **62**, 545–555 (2017)

47. Larena, I., De Cal, A., Linan, M., Melgarejo, P.: Drying of *Epicoccumnigrum* conidia for obtaining a shelf-stable biological product against brown rot disease. J. Appl. Microbiol. **94**, 508–514 (2003)

48. Sabuquillo, P., De Cal, A., Melgarejo, P.: Biocontrol of tomato wilt by *Penicillium oxalicum* formulations in different crop conditions. Biol. Control **37**, 256–265 (2006)

49. Rajonee, A.A., Zaman, S., Huq, S.M.I.: Preparation, characterization and evaluation of efficacy of phosphorus and potassium incorporated nano fertilizer. ANP. **6**, 1–13 (2017)

50. Chaudhary, R., Nawaz, K., Khan, A.K., Hano, C., Abbasi, B.H., Anjum, S.: An overview of the algae-mediated biosynthesis of nanoparticles and their biomedical applications. Biomolecules **10**, 1498 (2020)

51. Rastogi, A., Tripathi, D.K., Yadav, S., Chauhan, D.K., Zivcak, M., Ghorbanpour, M., El-Sheery, N.I., Brestic, M.: Application of silicon nanoparticles in agriculture. 3 Biotech **9**, 1–11 (2019)
52. Karunakaran, G., Suriyaprabha, R., Rajendran, V., Kannan, N.: Influence of ZrO_2, SiO_2, Al_2O_3 and TiO_2 nanoparticles on maize seed germination under different growth conditions. IET Nanobiotechnol. **10**, 171–177 (2016)
53. Prasad, T.N.V.K.V., Sudhakar, P., Sreenivasulu, Y., Latha, P., Munaswamy, V., Reddy, K.R., Pradeep, T.: Effect of nanoscale zinc oxide particles on the germination, growth and yield of peanut. J. Plant Nutr. **35**, 905–927 (2012)
54. Salama, H.M.: Effects of silver nanoparticles in some crop plants, common bean (*Phaseolus vulgaris*) and corn (*Zea mays*). Int. Res. J. Biotechnol. **3**, 190–197 (2012)
55. Berg, G., Alavi, M., Schmid, M., Hartmann, A.: The rhizosphere as a reservoir for opportunistic human pathogenic bacteria. Mol. Microb. Ecol. Rhizosphere **2**, 1209–1216 (2013)
56. Keswani, C., Prakash, O., Bharti, N., Vílchez, J.I., Sansinenea, E., Lally, R.D., et al.: Re-addressing the biosafety issues of plant growth promoting rhizobacteria. Sci. Total Environ. **690**, 841–852 (2019)
57. Martinez-Hidalgo, P., Maymon, M., Pule-Meulenberg, F., Hirsch, A.M.: Engineering root microbiomes for healthier crops and soils using beneficial, environmentally safe bacteria. Can. J. Microbiol. **65**, 91–104 (2018)
58. Mendes, R., Garbeva, P., Raaijmakers, J.M.: The rhizosphere microbiome: significance of plant beneficial, plant pathogenic, and human pathogenic microorganisms. FEMS Microbiol. Rev. **37**, 634–663 (2013)
59. Sessitsch, A., Pfaffenbichler, N., Mitter, B.: Microbiome applications from lab to field: facing complexity. Trends Plant Sci. **24**, 194–198 (2019)
60. Glick, B.R.: Introduction to plant growth-promoting bacteria. In: Beneficial Plant-Bacterial Interactions. Springer, Cham, pp. 1–37 (2020)
61. Mawarda, P.C., Le Roux, X., Dirk van Elsas, J., Salles, J.F.: Deliberate introduction of invisible invaders: a critical appraisal of the impact of microbial inoculants on soil microbial communities. Soil Biol. Biochem. **148**, 107874 (2020)
62. Hart, M.M., Antunes, P.M., Chaudhary, V.B., Abbott, L.K.: Fungal inoculants in the field: is the reward greater than the risk? Funct. Ecol. **32**, 126–135 (2018)
63. Ambrosini, A., de Souza, R., Passaglia, L.M.P.: Ecological role of bacterial inoculants and their potential impact on soil microbial diversity. Plant Soil **400**, 193–207 (2016)
64. Marschner, P., Timonen, S.: Interactions between plant species and mycorrhizal colonization on the bacterial community composition in the rhizosphere. Appl. Soil. Ecol. **28**, 23–36 (2005)
65. Di Salvo, L.P., Ferrando, L., Fernandez-Scavino, A., Garcia de Salamone, I.E.: Microorganisms reveal what plants do not: wheat growth and rhizosphere microbial communities after *Azospirillum brasilense* inoculation and nitrogen fertilization under field conditions. Plant Soil **424**, 405–417 (2018)
66. Eisenhauer, N., Schulz, W., Scheu, S., Jousset, A.: Niche dimensionality links biodiversity and invasibility of microbial communities. Funct. Ecol. **27**, 282–288 (2013)
67. Trabelsi, D., Mhamdi, R.: Microbial inoculants and their impact on soil microbial communities: a review. Biomed. Res. Int. **2013**, 863240 (2013)
68. Asiloglu, R., Shiroishi, K., Suzuki, K., Turgay, O.C., Murase, J., Harada, N.: Protist-enhanced survival of a plant growth promoting rhizobacteria, *Azospirillum* sp. B510, and the growth of rice (*Oryzasativa*) plants. Appl. Soil Ecol. **154**, 103599 (2020)
69. Mar Vazquez, M., Cesar, S., Azcon, R., Barea, J.M.: Interactions between arbuscular mycorrhizal fungi and other microbial inoculants (*Azospirillum, Pseudomonas, Trichoderma*) and their effects on microbial population and enzyme activities in the rhizosphere of maize plants. Appl. Soil Ecol. **15**, 261–272 (2000)
70. He, C., Wang, W., Hou, J.: Plant growth and soil microbial impacts of enhancing licorice with inoculating dark septate endophytes under drought stress. Front. Microbiol. **10**, 2277 (2019)
71. Monokrousos, N., Papatheodorou, E.M., Orfanoudakis, M., Jones, D.G., Scullion, J., Stamou, G.P.: The effects of plant type, AMF inoculation and water regime on rhizosphere microbial communities. Eur. J. Soil Sci. **71**, 265–278 (2020)

72. Ghosh, N.: Promoting biofertilizers in Indian agriculture. Econ. Pol. Wkly **39**, 5617–5625 (2004)
73. https://www.agrifarming.in/biofertilizer-subsidy-nabard-government-schemes (24 July 2022)
74. Lee, L.H., Wu, T.Y., Shak, K.P.Y., Lim, S.L., Ng, K.Y., Nguyen, M.N., Teoh, W.H.: Sustainable approach to biotransform industrial sludge into organic fertilizer via vermicomposting: a mini-review. J. Chem. Technol. Biotechnol. **93**, 925–935 (2018)
75. Khan, S., Ankit, P., Vadsaria, N.: Hydroponics: current and future state of the art in farming. J. Plant Nutr. **44**, 10 (2021)
76. Mohammad, A., Khan, A., Kuek, C.: Improved aeroponic culture of inocula of arbuscular mycorrhizal fungi. Mycorrhiza **9**, 337–339 (2000)

Chapter 14
Plant Tissue Culture Technique and Smart Agriculture

Zainab A. H. AL-Hussaini

Introduction

The world's total population is expected to reach up to 9.7 billion by 2050 and to feed such a huge number, food production rate must also increase by about 70%. Agriculture plays a significant role in meeting up these challenges. In order to attain a greater yield and productivity, it has to overcome various climatic changes and help mitigate its effects. Agriculture contributes to climate change either through retention of gas emissions from anthropogenic activities or through the conversion of non-agricultural lands such as forests into agricultural lands. Change in the use of forests as agricultural land has contributed about 20–25% of annual global emissions in 2010 [1].

Smart agriculture can be described as a method which is based on the use of advanced and sustainable agricultural technology, along with rationalizing the use of natural resources, especially water, and being reliant on information management and analysis system to make the best possible production outcomes, at lowest costs, and is characterized by a real possibility to provide a more productive and sustainable agro-production based on a more resource-efficient approach. Climate-smart agriculture is not a new set of sustainable production practices or systems, but rather is an approach aimed at providing means to integrate the specific characteristics of adaptation and mitigation in sustainable agricultural development policies, programs, and investments.

Z. A. H. AL-Hussaini (✉)
Presidency of the Molecular Biology Division Biotechnology—Plant Tissue Culture,
Ministry of Science and Technology, Baghdad, Iraq
e-mail: agrdg@mohesr.gov.iq

Department of Genetic Engineering, Agriculture Research Directorate, Ministry of Science and Technology, Baghdad, Iraq

© Centre for Science and Technology of the Non-aligned and Other
Developing Countries 2023
K. Pakeerathan (ed.), *Smart Agriculture for Developing Nations*, Advanced Technologies and Societal Change, https://doi.org/10.1007/978-981-19-8738-0_14

Smart Agriculture at the Arab Level

The Arab is among the regions which are facing drastic environmental crisis, such as lack of arable water for crop growth, climatic changes, drought, and desertification, adversely affecting the provision for food and food security. Hence, the Arab region is in an ultimate need for the application of smart farming techniques. However, there are few obstacles in incorporating this method like weak communication and Internet infrastructure in some Arab countries, lack of technical skills, and high material cost. Therefore, adopting smart agriculture policies require promoting Joint Arab action, exchange of knowledge and ideas about Internet of Things technology, and integrating information and communication technology as an essential factor for sustainable agricultural development [2].

Plant Tissue Culture

The plant cell and tissue culture technique are one of the important techniques of biotechnology that challenges the problems related to climate change. It is considered to be one of the latest, significant, and remarkable disciplines in advanced agricultural sciences, which are currently being used and have provided many beneficiary uses. They are neither directly nor indirectly affected by climatic changes, which cast adverse effects on agricultural production and yield in all countries of the world. This culture technique has many advantages. In general, it is not affected by the harmful effects of climate changes. By adopting cultivation within a small controlled area, with high productivity and resistance to bad and difficult environmental conditions, as well as cultivation throughout the year ignoring rainy season or the appropriate season for production, and the speed of production, it is possible to obtain a high yield of productivity within a short period of time, ensuring that they are disease-free and can be exported to all countries of the world, as well as obtaining an environment devoid of solid, liquid, or gaseous waste that is harmful to the environment, humans, or any other life forms.

Recently, attention has been turned toward the "Agriculture Technology" in balancing the natural equation and taming the desert to find sustainable solutions for the development of agriculture, including the use of molecular markers, artificial insemination, biofertilizers, micro propagation, genomics, and genetic modifications. There are achievements made in this field in some Arab countries. For example, the plant tissue culture laboratory in Al Foah area in Al Ain city, affiliated to the United Arab Emirates University, is one of the largest laboratories both regionally and globally, involved in commercial palm propagation. The laboratory produces 40–50 thousand advanced high-quality palm plantlets annually, by using the "organogenesis" technology, which is one of the best scientific techniques for the production of palm tissue plantlets resistant to palm weevil and tolerant of high salinity and drought [3].

The Challenge

Biotic and abiotic environmental stresses, especially drought and salinity possess major threat to agricultural activities, thus causing reduction in plant growth and crop yield.

In Iraq, the potato crop is considered as a widely used consumer crop. Statistics indicate that potato cultivation in the world reached 368 million tons on an area of 17.5 million hectares, unlike Iraq producing about 294 thousand tons on an area of 18,780 ha [4]. Hence, it was noted that this production was insufficient to meet up the current demands of the country which necessitated the need for importing it from the Arab and international markets to fulfill the deficit caused.

Concept of Project Idea

Expanding genetic variations to obtain tolerant genotypes can be achieved by employing irradiation techniques (physical or chemical mutagens) with plant tissue culture (somaclonal variation) providing a good opportunity in obtaining salt-tolerant clones. Somaclonal variation in combination with in vitro mutagenesis is efficient for exploiting and isolating salt-tolerant cell lines in a short duration speeding up the creation of variation.

The Protocol

A protocol for induction of mutant clones [plants induced from salt-tolerant calli (planted at salt levels 8, 10, 12 dsm^{-1}) and plants induced from non-salt-tolerant calli (planted at salt level 6 dsm^{-1})] and their parental cultivars of potato crop have been developed through in vitro, pre-selected by exposing plants primarily to saline stress conditions, and then planting them in the field to ensure the trait is inherited in the progeny. Studies are going since several generations to ensure that salt tolerance is consistent [5].

References

1. The enterpriseworld: Top 5 Modern Agriculture Technologies that Made Farming Smarter. https://theenterpriseworld.com/modern-agriculture-technology
2. Climate-Smart Agriculture: imperative practices for adapting to climate change. https://www.scientificamerican.com/arabic/articles
3. Khlifa, M.: Taming the Desert with "Agriculture Technology" (2022). https://www.albayan.ae
4. FAOSTAT (2020)
5. AL-Hussaini, A.A.H.: *In Vitro* employment of genetic variation in potato (*Solanum tuberosum* L.) to Improve salt tolerance. Ph.D. Thesis., Faculty of Agriculture, Kufa University, Iraq (2016)

Chapter 15
Investigation of Benzene Residue Removal from Agricultural Soil After Gamma Irradiation Treatment Using GCMS and FTIR Method

Moe Min Htwe and Khin Myo Sett

Introduction

Soil changes, including formation, degradation, the outcome of agricultural use, anthropogenesis, remediation, and environmental impacts, require an increasing volume and quality of information [1, 2]. Agriculture is an important part of the Myanmar economy, and pesticide poisoning is a potential agricultural hazard. Pesticides are used for crop yield increasing and crop protection but its residues affect in different environmental compartments, e.g., soil, water, fruits, and vegetables have adverse effects on the human health, and may be leached to the groundwater. There may be plenty of pesticide left in cultivated soil and may remain toxic and others that can affect the soil quality and fertility [3]. Fourier-transform infrared (FTIR) spectroscopy is a unique tool for the study of mineral and organic components of soil samples. FTIR spectroscopy offers sensitive characterization of minerals and soil organic matter (SOM), and mechanistic and kinetic aspects of mineral–SOM interactions that underlie biogeochemical processes. The versatility of FTIR spectroscopy makes it a foundational tool for soil scientists, despite challenges in the acquisition and interpretation of soil spectra that stem from chemical heterogeneity [4]. GC–MS can be used to separate complex mixtures, quantify analytes, identify unknown peaks and determine trace levels of contamination when combined with the detection power of mass spectrometry (MS). Pesticides constitute a very important group of chemical compounds that have to be controlled due to their high toxicity and their widespread use in agricultural practice for field and postharvest protection. The presence of pesticide residues in food is a direct result of pesticide use on crops. Over 1000 compounds may be applied to agricultural crops in order to

M. M. Htwe (✉) · K. M. Sett
Department of Atomic Energy, Ministry of Science and Technology, Nay Pyi Taw, Myanmar
e-mail: most18.nayitaw@gmail.com

© Centre for Science and Technology of the Non-aligned and Other Developing Countries 2023
K. Pakeerathan (ed.), *Smart Agriculture for Developing Nations*, Advanced Technologies and Societal Change, https://doi.org/10.1007/978-981-19-8738-0_15

control undesirable molds, insects, and weeds [5]. Soil samples can be analyzed by FTIR spectroscopy using a variety of methods, the most common of which are transmission, diffuse reflectance infrared Fourier-transform spectroscopy (DRIFTS), and attenuated total reflectance (ATR). Different modes of acquiring FTIR spectra offer complementary methods for evaluating soil components and processes. According to these facts, there is need to check the screening analysis and quantitative analysis of pesticide in agricultural soil which was planted vegetables. Pesticides are applied to crops to increase their yield. However, pesticide residues in different environmental compartments, e.g., soil, water, fruits, and vegetables have adverse effects on the human health. In this respect, continuous monitoring of pesticides is required in different environmental matrices at low concentration levels.

Materials and Methods

Sample Collection and Investigation

Soil sample collection was performed according to standard guideline in rainy season. Soil samples are collected from two agricultural lands in Mandalay region which were cultivated last year.

Real soil samples were collected in fields which have cultivated and are located near Mandalay Region, Myanmar. The samples were taken in pumpkin fields. Soil samples were collected in ten points from one acre among 10 acres of cultivated land in zigzag pattern according to soil sampling guideline with google map location application. Samples are collected at depth (15–25 cm). Investigation of harmful organic pollutants will be performed before and after irradiation treatment of collected soil samples. In this work, screening test for samples in three fields is to be performed. These samples from ten points of each field were mixed thoroughly, sieved and stored in air tight zipped plastic bag at 4 °C until the soil sample would be analyzed.

The soil sample was extracted with sodium chloride, methanol, and dichloromethane until 1 ml using air-dry oven, and 1 ml normal hexane was added; and after that, the extract was filtered using 0.4 μm filter paper and then injected for GCMS measurement. The analysis of pesticides was performed by using of high-performance Clarus® Gas Chromatograph/Mass Spectrometers. FTIR analysis was done for agricultural soil using MODEL Shimadzu IR Prestige-21 FTIR.

Gamma Irradiation Treatment

All agricultural soil samples are treated with gamma radiation (3 kg, 5 kGy, 10 kGy, 15 kGy, and 20 kGy) for investigation of degradation of pesticide by using Gamma Chamber GC 5000 with activity is 12 kCi (since 2000), and the current dose rate is ~ 0.67 kGy/hr, Total of 30 samples for two analyses were irradiated from two different fields.

Investigations by GCMS and FTIR Method

The investigations were performed for screening of pesticide residue from agricultural soil samples in Kyaukse district, Mandalay Division by using GCMS analysis and FTIR analysis. As first investigation of pesticide residues percentage in samples in same field after one year for comparison of pesticide residue content after gamma irradiation treatment was performed by using GCMS and FTIR analysis. As second investigation, gamma irradiation treatment of soil samples in two agricultural fields which were collected this year using GCMS and FTIR spectroscopy.

Results and Discussion

For investigation, there were two analyses for pesticide residue percentage;

1. Investigation of pesticide residues percentage changes in samples from same field after one year for comparison of pesticide residue content before and after gamma-irradiation treatment using GCMS and FTIR spectroscopy.
2. Investigation of gamma irradiation treatment of soil samples in two agricultural fields which were collected this year using GCMS and FTIR spectroscopy.

For first analysis (field 1 and field 2), field 1 samples were collected last year and field 2 samples were collected this year. The screening the pesticide residue in the soil samples (field 1 and field 2) was investigated by GCMS method before and after gamma irradiation treatment (3 kg, 5 kGy, 10 kGy, 15 kGy, 20 kGy). From these GCMS chromatograph benzene residues (Benzenebutanoic acid, Benzene, 1, 3-Dimethyl benzofuran)-still exists in field 1 before gamma treatment. After gamma irradiation treatment above 10 kGy, concentration of benzene residue reduced to around 80%.

For second analysis (field 2 and field 3), agricultural soil samples from two different fields were collected in same day and treated with gamma radiation (3 kg, 5 kGy, 10 kGy, 15 kGy, 20 kGy) for investigation of degradation of pesticide. The determination of pesticide residue was performed by GCMS from two agricultural farm lands in same region, Kyaukse Township. In both fields, there were Benzene residues (O-methoxybenzonitrile, Benzene, 1-isocyano-2-methoxyethylbenzene, Benzenethanol, Alpha, beta-dimethyl) in soil. After application of 10 kGy dose and above, the mentioned benzene groups decreased 30% in comparison with unirradiated samples in GCMS analysis. Efficient removal of benzene and chlorinated benzenes in effluents is of importance for environmental protection [6].

In the FTIR spectrum of benzene of field 1, filed 2, and field 3, peaks detected between 3500 and 1000 cm^{-1} due to aromatic stretching of–CH. After irradiation, transmission peaks decreased up to 30–35% depending on doses. From these screening results, benzene and organophosphate group decreased from 10 kGy dosage. FTIR results were consistent with the GCMS results as benzene residue existed in all samples. These results can support for degradation of organic pollutant using gamma irradiation. The quantitative results of benzene residues will be determined by GCMS method to ensure the accurate results of the occurrence of emerging organic pollutants in agricultural soil. The radiolytic degradation of halogenated alkyl and aromatic hydrocarbons for environmental purposes has been the subject of several papers; Hashimoto et al. demonstrated the efficient degradation of hydroxybenzenes and chlorophenols [7] (Figs. 15.1 and 15.2).

Conclusion

Benzene concentrations of 2.5, 14, and 250 ppm are acceptable for residential, industrial, and recreational soils, respectively. In these measurements of GCMS and FTIR screening agricultural soil, benzene residue was determined after gamma irradiation treatment. 10 kGy, concentration of benzene residue reduced to around 80%. The quantitative measurement for benzene residue and organophosphate residue should be performed for the recommendation.

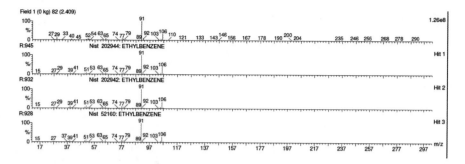

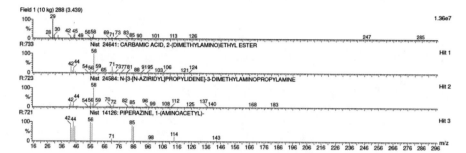

(a) GCMS chromatograph of field 1 before and after gamma irradiation treatment

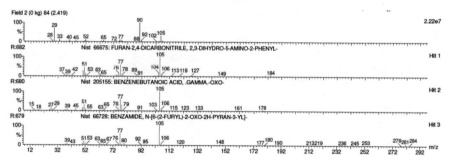

Fig. 15.1 GCMS chromatograph of field 1, field 2, and field 3 before and after gamma irradiation treatment

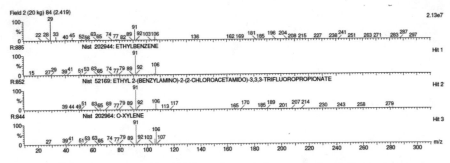

(b) GCMS chromatograph of field 2 before and after gamma irradiation treatment

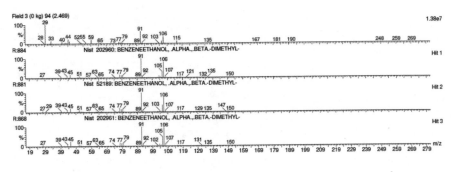

(c) GCMS chromatograph of field 3 gamma irradiation treatment

Fig. 15.1 (continued)

Acknowledgements The authors are grateful to the authorities: Ministry of Science and Technology, Myanmar and International Atomic Energy Agency, IAEA, participating in Coordinated Research Project (CRP) title 'Radiation based technologies for treatment of emerging organic pollutants. The award of IAEA, CRP project is also acknowledged. IAEA CRP code. F23034.

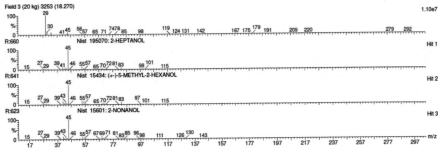

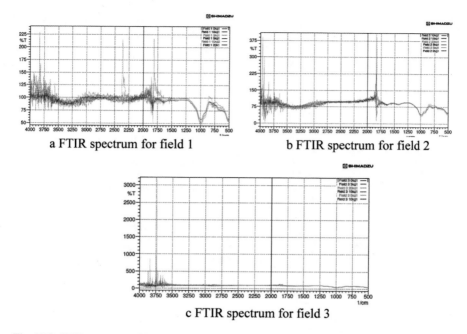

a FTIR spectrum for field 1

b FTIR spectrum for field 2

c FTIR spectrum for field 3

Fig. 15.2 FTIR spectrum of field1, field 2, and filed 3 before and after gamma irradiation treatment

References

1. Ichwana, I., Nasution, Z., Munawar, A.A.: The application of fourier transform infrared photoa-coustics spectroscopy (FTIRPAS) for rapid soil quality evaluation. Rona Tek. Pertan. **1**, 1–10 (2017)
2. Volkov, D.S., Rogova, O.B., Proskurnin, M.A.: Photoacoustic and photothermal methods in spectroscopy and characterization of soils and soil organic matter. Photoacoustics **17**, 100151 (2020). https://doi.org/10.1016/j.pacs.2019.100151
3. Semra, K., Tuncel, G.: Study of pesticide contamination in soil, water and produce using gas chromatography mass spectrometry. J. Anal. Bioanal. Tech. **9**(4), 409 (2018)
4. Margenot, A.J., Calderón, F.J., Goyne, K.W., Mukome, F.N.D., Parikh, S.J.: IR spectroscopy, soil analysis applications. In: Lindon, J.C., Tranter, G.E., Koppenaal, D.W. (eds.) The Encyclopedia of Spectroscopy and Spectrometry, vol. 2, 3rd edn., pp. 448–454. Academic Press, Oxford (2017)
5. Asep Bayu Dani Nandiyanto, B., Rosi, O., Risti, R.: How to read and interpret FTIR spectroscope of organic material. Indonesian J. Sci. Technol. **4**(1), 97–118 (2019)
6. Trojanowicz, M., Chudziak, A., Bryl-Sandelewska, T.: Use of reversed-phase HPLC with solid-phase extraction for monitoring of radiolytic degradation of chlorophenols for environmental protection. J. Radioanal. Nucl. Chem. **224**(1–2), 131–136 (1997)
7. Nickelsen, M.G., Cooper, W.J., Waite, T.D., Kuruzc, C.N.: Removal of benzene and selected alkyl-substituted benzenes from aqueous solution utilizing continuous highenergy electron irradiation. Environ. Sci. Technol. **26**, 144–152 (1992)

Chapter 16
Removal of Nitrates from Drinking Water Using Chia Seeds in Gaza Strip, Palestine

Salah El Sadi, Hassan Tammous, and Khamis Al mahllawi

Introduction

One of the most dangerous issues of water contamination is the nitrate anion. As a natural source, nitrate present in moderate concentrations, which resulted from the degradation of organic nitrogen-containing compounds. Their presence in nature is mostly in rocks, organic matter and soil. The presence of decaying organic matter deeply in the earth is a polluting source of nitrates in nature [6].

Human resource nitrate levels in groundwater keep on rising. The major cause of such levels of nitrate is the increased use of fertilizers and pesticides as well as the filtration of wastewater into the aquifer. Numerous variables can influence nitrate groundwater contamination levels, like fertilizers, fertilizer utilization rate, crop administration and the sort of nitrogen utilized [15] and also, as a human resources, chemicals introduced in the manufacture of explosives and unloading them as waste materials. Additionally, from sources cultivation of leguminous crops acting on the stabilization of nitrogen in the atmosphere which rise nitrates concentration [1].

The negative activities that cause the degradation of fresh water can be categorized into five types, such as point contamination sources, diffuse contamination sources, groundwater over exploitation, artificial recharge and seawater intrusion

S. El Sadi (✉)
Department of Water and Environment, Ministry of Economy, Gaza, Palestine
e-mail: salah_sady@hotmail.com

H. Tammous · K. Al mahllawi
Department of Chemistry, Faculty of Science, Al Azhar University, Gaza, Palestine
e-mail: dr.hassantamous@gmail.com

K. Al mahllawi
e-mail: khamis271066@gmail.com

K. Al mahllawi
Department of Environment and Water Quality, Environment Quality Authority, Gaza, Palestine

© Centre for Science and Technology of the Non-aligned and Other
Developing Countries 2023
K. Pakeerathan (ed.), *Smart Agriculture for Developing Nations*, Advanced Technologies and Societal Change, https://doi.org/10.1007/978-981-19-8738-0_16

[10]. Among them, diffuse contamination due to human activities is a major cause of water pollution. Nitrate pollution of water due to intensive agricultural activities has become a major environmental problem since 1970s. Nitrate is highly soluble in water and does not readily bind to the soil causing it to be highly susceptible to leaching. There are several potential sources of nitrate, including animal wastes, septic tanks, and municipal wastewater treatment systems and decaying plant debris. However, nitrogen enriched fertilizers for farming is considered as the main source of nitrate pollution in the environment. When nitrates form in the water at high concentrations, nutrients play a rapid role in the growth of algae in water; algae consume dissolved oxygen gas in the water, resulting in the suffocation of fish and leads to the death of fish [14]. Therefore, the magnitude of these health risks should pay attention to the human body and protect it, noting that nitrates should not exceed 50 mg/L in drinking water, according to the World Health Organization to protect human health from risk [18]. It was necessary to find solutions to address the risk of these nitrates. Many scientific researches have been conducted to treat the nitrate ion to provide an appropriate solution for these ions using inexpensive economical methods. New systems, such as biological denitrification removal systems, have been used as an example of anoxic microbial processes, which have been used as a way of dealing with the risks of this ion. This operation can convert nitrate into nitrogen in four enzymatic steps via the following intermediates: nitrite (NO^{2-}), nitrogen oxide (NO) and dinitrogen oxide (N_2O) [17]. The utility of these processes has been limited due to their expensive operation and subsequent disposal problem of the generated nitrate waste brine [16]. The present study aimed to investigate the efficiency of Chia seeds for removal of nitrates from drinking water by analyzing several parameters.

Materials and Methods

Chia Seeds Biosorbent Preparation

Preparation of Chia seeds biosorbent was carried out at various main stages which include: collection from the relevant places, burning, washing thoroughly, air drying, grinding, sieving and finally drying in ovens at different temperatures. The prepared Chia seeds were stored in tight containers for the experimental testing of nitrate removal at various factors.

Collection of Chia Seeds

Chia seeds were collected freshly from the distribution points of Chia local production factories.

Burning of Chia Seeds

After collecting Chia seeds, the Chia is placed in a crucible and then burned inside the oven at a temperature of 650 °C for 30 min.

Chia Seeds Washing After Burning

After washing the Chia seeds to ensure the removal of contaminants, Chia seeds were air dried for 12 h at different temperatures (20–80 °C) for 24 h.

Chia Seeds Sieving

After drying of the Chia seeds, the Chia seeds were grinded using pestle and mortar at suitable particle sizes. The grinded Chia seeds were sieved at different particle sizes using standard numbered sieves. Three types of particles with different size ranges were collected (coarse = 2.8 mm–710 μm, medium = 710 μm–90 μm and fine < 90 μm).

Chia Seeds Weighing

Samples of Chia seeds of different particle sizes were weighed at variable weights. The weights tested were 1, 2 and 5 g.

Preparation of Nitrate Solutions (Adsorbate).

Stock nitrate solutions ($NO_3^-{}_{aq}$) (1000–1500 mg/L) were prepared using distilled water from a pure potassium nitrate in volumetric flasks and prevented from direct sunlight by covering with aluminum foils and stored in a refrigerator. Dilute solutions of the desired concentrations (50–1500 mg/L) were prepared in volumetric flasks by dilution using distilled water when required.

Preparation of Acetate and Citric Solutions for pH Control

Different pH values (3.5–10) were controlled using acetic acid (0.1 M) and Citric acid.

Nitrate Removal Experiments by Chia Seeds Biosorbent by Batch Method

In the batch method, a determined amount of Chia seeds biosorbent (adsorbent) of known particle size was mixed in a specific volume of nitrate solution (adsorbate) with a definite concentration. The mixture was kept at variable factors which are discussed in the coming sections in order to examine their effect on the removal of nitrate from the lab prepared nitrate aqueous solutions and the natural ground water. After monitoring all factors, the optimum conditions for the highest rate removal of maximum nitrate amount were determined and controlled for further testing. These factors include: temperature of Chia burning, temperature of nitrate solution (adsorbate) during mixing with Chia seeds, contact time, particle size of Chia seeds adsorbent, nitrate concentration, pH value and the recovery of Chia seeds adsorbent. Each experiment was conducted in triplicates to obtain the mean value. Control experiments performed without addition of adsorbent' confirmed that self-degradation of nitrate was negligible. The amount of nitrate removed per unit Chia seeds adsorbent (mg nitrate per g adsorbent) was calculated at time t by Eq. (16.1) and at equilibrium by Eq. (16.2).

$$q_t = \frac{(C_i - C_t)V}{W} \tag{16.1}$$

$$q_e = \frac{(C_i - C_e)V}{W} \tag{16.2}$$

where q_t and q_e are the amount of nitrate removed in (mg/g) at time t and at equilibrium, respectively; C_i, C_t and C_e are the nitrate concentrations at zero time, at time t and at equilibrium in (mg/L), respectively; V is the solution volume in (mL); and W is the sorbent dosage in (g).

The efficiency of nitrate removal at time t and at equilibrium, respectively, was determined using Eqs. (16.3) and (16.4):

$$\% \text{ Removal at time } t = \frac{(C_i - C_t) \times 100}{C_i} \tag{16.3}$$

$$\% \text{ Removal at equilibrium} = \frac{(C_i - C_e) \times 100}{C_i} \tag{16.4}$$

where C_i, C_t and C_e are the nitrate concentrations at zero time, at time t and at equilibrium in (mg/L), respectively.

Factors Affecting Absorption Processes

Effect of Contact Time

The percentage of nitrate removal from aqueous solutions was determined by mixing a given amount of Chia seeds biosorbent (1–5 g) with 100 mL of 100 mg/L concentration of nitrate ($NO_3^-{}_{aq}$) solution using 100 mL Erlenmeyer flask at room temperature. Measurements of the amount of nitrates removed were carried out at different time intervals hours and days. At each time, a definite volume of the solution was withdrawn by a micropipette and diluted with a suitable amount of distilled water to the linear range of nitrate calibration curve that is based on five standard solutions of nitrate at the concentration range (5–20 mg/L). The diluted amount was filtered and the filtrate was analyzed for nitrate using spectrophotometer at solution at the characteristics wavelength ($\lambda_{max} = 220$ nm). A controlled nitrate sample of (100 mL, 100 mg/L) without the addition of Chia seeds material was used as a blank. The removal efficiency was determined as % nitrate and as mg/g Chia seeds biosorbent material [19].

Effect of Oven Temperature

The effect of Oven temperature of Chia seeds was examined at different temperatures (200, 350, 450, 550, 650 and 900 °C). The experiments were conducted at nitrate solution volume (100 mL), nitrate initial concentration (100 mg/L), and pH value (5.5–8.5). The amount of nitrate removed as % nitrate removal efficiency was determined at different time intervals within four days. The % of nitrate removed was determined versus different temperatures [5].

Effect of Nitrate Initial Concentration

Different initial concentrations of nitrate aqueous solutions (50, 100, 250, 500, 1000 and 1500 mg/L) of a volume of 100 mL were mixed individually with a given amount of oven dried Chia seeds biosorbent (5 g) using 100 mL Erlenmeyer flask at room temperature. Measurements of the amount of nitrate removed were carried out at different time intervals. The amount of nitrate removed was determined versus concentration as mg/g Chia seeds biosorbent material.

Effect of Particle Size

The effect of adsorbent particle size on nitrate removal was studied preliminary using different particle sizes (coarse = 2.8 mm–710 μm, medium = 710 μm–90 μm and fine < 90 μm) by maintaining the pH (7–7.5), initial nitrate concentration (100 mg/L), volume of adsorbate (100 mL) and adsorbent dosage (5 g). The experiments were conducted at temperature of both the biosorbent and adsorbate solution at 20 °C, respectively. The amount of nitrate removed as % nitrate removal efficiency was determined at different time intervals within four days. Further experiments for examining the effect of particle size of the adsorbent were conducted by modification of the conditions in order to improve the % removal efficiency of nitrate by the adsorbent.

Effect of Adsorbent Dose

The effect of amount of adsorbent (dose) on nitrate removal was studied using different doses of Chia seeds adsorbent (1 and 10 5). The experiments were conducted at medium particle size of the adsorbent (710 μm–90 μm), nitrate solution volume (100 mL), nitrate initial concentration (100 mg/L), and pH value (5.5–8.5). The amount of nitrate removed as % nitrate removal efficiency was determined at different time intervals within four days. The % of nitrate removed was determined versus adsorbent dose.

Effect of Solution Temperature

The effect of temperature of Chia seeds adsorbent mixed with nitrate solutions either Stock or ground water samples were examined at different temperatures (20, 40, 60, 80 and 100 °C). The experiments were conducted at particle size of the adsorbent at (coarse = 2.8 mm–710 μm), nitrate solution volume (100 mL), nitrate initial concentration (100 mg/L), and pH value (5.5–8.5). The amount of nitrate removed as % nitrate removal efficiency was determined at different time intervals within four days. The % of nitrate removed was determined versus drying temperature.

Effect of pH

The effect of pH on amount of nitrate removal was analyzed over a pH range of (3.5–10.5). The pH was adjusted using 0.1 N acetic acid and 0.1 N sodium acetate solutions and measured by a pH meter (AD 1020). In this study, experiments were carried at medium particle size of the adsorbent (710 μm–90 μm), nitrate solution volume (100 mL), nitrate initial concentration (100 mg/L) and the amount of adsorbent dose (5 g). The amount of nitrate removed as % nitrate removal efficiency was determined at different time intervals within four days. The % of nitrate removed was determined versus pH value [13].

Optimization Conditions

After a number of testing experiments where the various factors were applied for the % removal of nitrate the researcher applied the best conditions that gave maximum removal of nitrate especially in real ground water samples. These factors include: amount of adsorbent dosage (5 g), pH (5.5), contact time (3 days), temperature (25 °C), particle size (2.8 mm), adsorbent temperature (20 °C) and finally adsorbate temperature (25 °C).

Applying the Chia Seeds Biosorbent for Nitrates Removal from Groundwater

After several developments and changing of factors to reach the best removal of synthetic nitrate solution, real groundwater samples were tested for nitrate in Gaza Strip at the optimum conditions. A volume of 100 mL of the ground water was mixed with 5 g of the Chia seeds biosorbent the optimum factors were controlled. After 48 h the amount of nitrate removed was determined. The test was repeated three times.

Recovery of Chia Seeds

The recovery procedure of Chia seeds biosorbent after its use for nitrate removal was conducted by testing the Chia seeds material in repeated experiments at the optimum conditions. Tests were carried out either after washing of the previously used material or without washing. Each experiment was carried out for 4 times at 48 h intervals at each one.

Detection of NO_2^-, −N (Nitrogen, Nitrite)

In order to detect if any nitrate converted to nitrite during the removal of nitrate by the Chia seeds biosorbent, nitrite test was conducted by the sulphanilamide spectrophotometric standard method [11]. Reagent for nitrite analysis was prepared (Scheme 16.1). Phosphoric (50 mL, 85%) acid was dissolved in 200 mL distilled water followed by 5 g of sulphanilamide, and then 0.5 g of N-(1-naphthyl) ethylenediamine was added to the mixture with stirring. The solution mixture was then dissolved to 500 mL by distilled water.

Because of ion instability, the samples of nitrite should be analyzed immediately after collection. Nitrite sample (1 ml) was withdrawn using a micropipette then 1 ml of nitrite reagent was mixed for 1 min until pink color appears. All samples

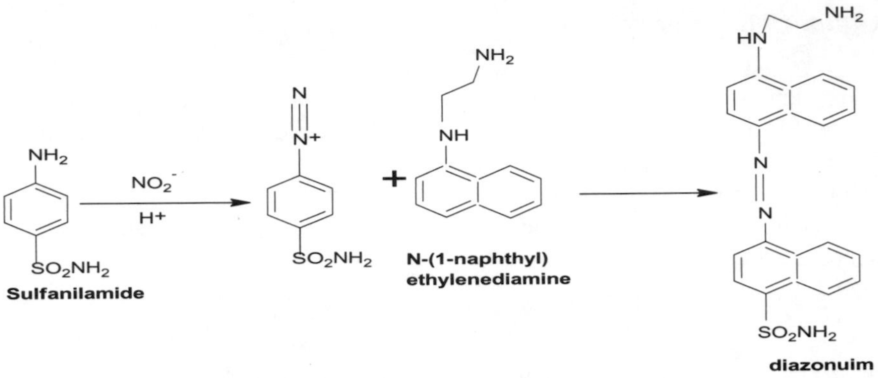

Scheme 16.1 Reagent for nitrite analysis [7]

were determined at wavelength 543 nm at spectrophotometer by applying calibration curve.

Detection of NH₃ –N, (Nitrogen, Ammonia)

Detection of NH$_3$ –N, (Nitrogen, Ammonia)

Ammonia is a species that could be produced as a byproduct during denitrification. Detection of ammonia was carried out by distillation and analyzed titrimetrically by Kjeldahl method. A sample (100 mL) was distilled and about 50 mL of distillate was collected into a solution of boric acid (20 mL) as absorbent solution. Few drops of bromocresol green and methyl red indicators mixture was added. The color changed from pink to green. The ammonia was determined by back titration using HCl (0.1 N) until green color disappeared. A blank sample was prepared. The volume of HCl consumed was determined and the amount of ammonia was calculated by the Eq. (16.5) [7]

$$NH_3 \text{--} N \frac{mg}{1} = \frac{(T - B) \times N \times 17.004 \times 1000}{C} \tag{16.5}$$

where

T Volume of HCl solution consumed (ml) for sample
B Volume of HCl solution consumed (ml) for blank
N Normality of HCl
C Sample volume.

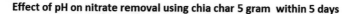

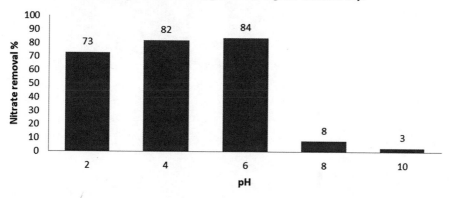

Fig. 16.1 Relation between pH and Chia seed after treatment as charcoal to nitrate removal in synthetic water, nitrate 100 mg/l

Result and Discussion

Effect of pH on % Nitrate Removal by Chia Seeds Char

Different pH values were used to examine the effect of pH on % nitrate removal by Chia seeds char. The various pH values (2, 4, 6, 8 and 10) were controlled using acetic acid and sodium hydroxide solution. The 100 mL of water solution of initial nitrate concentration 100 mg/L was mixed with 5 g Chia seeds char and the amount of nitrate removed was tested. The results are shown in Fig. 16.1. It is observed that the amount of nitrate removed increases with decreasing pH value and reaches the maximum at pH 4, 6 and 8 after 5 days contact time.

Effect of Chia Seeds Char Dose on % Removal of Nitrate

The effect of absorbent dose on the % efficiency of nitrate removal was studied. Different doses of Chia char material were used (1–5 g) that of particle size (2 mm). The nitrate solution (100 mL) was mixed with the Chia seeds char for 5 days. Determination of nitrate absorbed where conducted at different time intervals. The results are shown in Fig. 16.2; it is observed that at the first three days the % of nitrate removed increases with increasing of absorbent dose. Such a trend is mostly attributed to an increase in the sorption surface area and the availability of more active adsorption sites at higher amount of adsorbent [8]. After five days equilibrium attained and the % of nitrate removal is increase of the larger dose (86% for 5 g dose) comparing with both the smaller doses (79% for both 3–4 g doses) which attributed to the availability of the low doses to absorb more nitrate at long time.

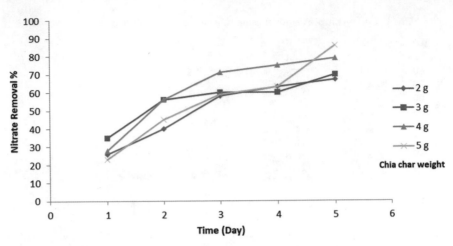

Fig. 16.2 Relation between Chia seeds char dosage (g) and nitrate percent removal with 2 ml acetic acid

Comparison Between Chia Char with Citric 0.2 M and Acetic Acid 0.2 M for Nitrate Removal from Synthetic Water 100 mg/l Nitrate (days)

In this experiment, a sample of Chia 5 g added with 2 mL of acetic acid was compared with another sample of Chia 5 g with 2 mL of citric acid in 100 ml of synthetic water 100 mg/l nitrate, these experiments were conducted for each sample at pH 5.8, 25c for 5 days, the result showed that the best removal was with citric acid more than acetic acid where nitrate removal percent with citric acid was 96% but 72% with acetic acid, this refer to the best removal with citric acid [9]. As Fig. 16.3

Effect of Particle Size of Chia Char on % of Nitrate Removal

The particle size is an important parameter owing to its effect on % removal efficiency and on the amount of nitrate adsorbed per unit weight of Chia chars. The percentage of nitrate removal from aqueous solution by Chia char was examined using 5 g of the Chia char at different particle sizes (2 mm, 1.18 mm, 600 μm and 425 μm). The pH was controlled at 5.8. The nitrate solution (100 mL, 100 ppm) was mixed with the Chia char and the amount of nitrate removed was determined. The results are shown in Fig. 16.4 it is observed that the amount of nitrate removed increases with 2 mm of particle sizes and reaches the minimum range at 425 particle sizes.

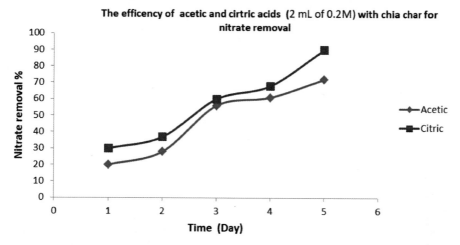

Fig. 16.3 Comparison between Chia char (5 g) with addition of 2 ml of 0.2 M citric and acetic acid on nitrate removal

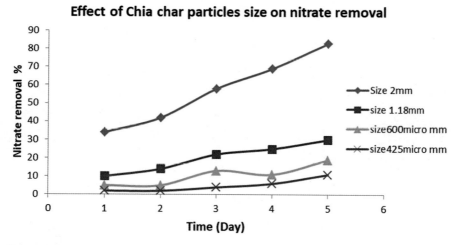

Fig. 16.4 Effect of particle size of Chia char on % of nitrate removal

Chia Char Recovery

Exhausted Chia char was re-tested for nitrate removal at the optimum conditions with or without reconditioning. Four cycles measurements tested Chia char with and without washing by distilled water. The cycles were repeated for four cycles and tested for (3 to 5 days) to check the recovery activity of the adsorbate. The results are given in Figs. 16.5 and 16.6, and we observed that the Chia char preserved its activity toward the removing of nitrate from water samples. The results show that the removal efficiency is 78–75% in case of unwashed samples while the removal

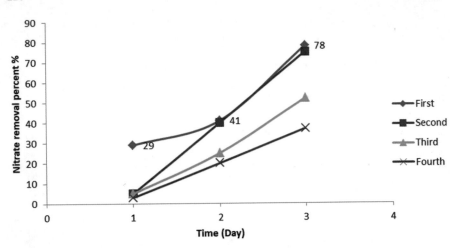

Fig. 16.5 Chia recycling without washing in other synthetic water sample 100 mg/L

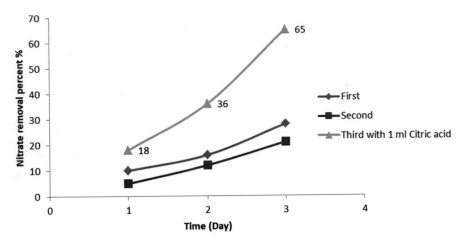

Fig. 16.6 Chia recycling with washing in other synthetic water sample 100 ppm

efficiency is 21–28% in case of washed samples. This attributed of the presence of acetic or citric acids that sustained its activity onto the Chia char adsorbate.

The Effects of Nitrate Initial Concentration on Chia Char Removal Efficiency for Nitrate

In this section, different concentrations of nitrates (100, 250, 500, 1000, 1500 mg/L) were prepared and 5 g of Chia char was placed in 5 batches (100 ml) with different

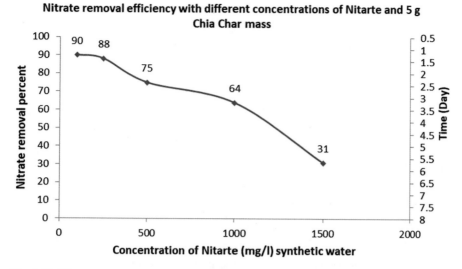

Fig. 16.7 Nitrate removal efficiency with different concentration of Nitrate

nitrate concentration and the remaining variables, i.e., pH, contact time and dosage of adsorbent were fixed. The result showed that the higher the nitrate concentration the less the nitrate removal efficiency within 5 days' time period as shown in Fig. 16.7, but after increasing the amount of chia from 5 to 15 g there was a change in the percentage of nitrate removal as shown in Fig. 16.8. It is observed that percent of nitrate removal increase with increase of initial concentration. The removal efficiency was better and removal time becomes less but with very high nitrates concentrations the removal rate was decreased. Effect of initial concentration of nitrate on percentage of nitrate ion removal from solution using Chia char is shown in Figs. 16.7 and 16.8.

Optimum Conditions of Chia Char for the % Removal Efficiency of Nitrate

The improvement of the efficiency of the Chia char for removing nitrate from aqueous solutions from the previous results could be summarized in Table 16.1.

Effect of Contact Time and Initial Nitrate Concentration on Percentage Nitrate Removal by Chia Seeds Biosorbent

Investigation the effect of contact time required to reach equilibrium is essential for designing batch adsorption experiments. Preliminary experiments were conducted

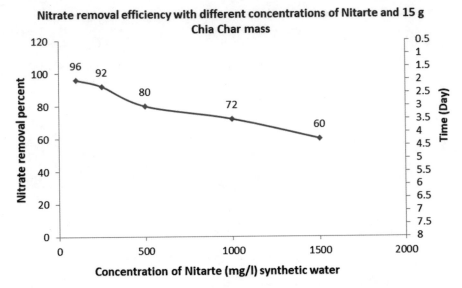

Fig. 16.8 The effect of Chia dose on nitrate removal efficiency with different concentration of nitrate

Table 16.1 Optimum conditions of Chia Char for nitrate removal efficiency

1	Dose	5 g and greater
2	pH	5.8–6.0
3	Time	5 days
4	Oven temperature (burning)	650–750 °C
5	Temperature	25 °C
6	Particle size	2 mm
7	Removal of nitrate (%)	90

to determine the percentage of nitrate removal from aqueous solution by Chia Seeds biosorbent at different concentrations (50, 100, 250, 500, 1000 and 1500 ppm) under room temperature using 5–15 g of the adsorbent powder of particle size (2 mm, 1.18 mm, 600 μm and 425 μm) which dried at 80 °C. Each nitrate solution (100 mL) was mixed with the Chia seeds substrate for five days. Determination of nitrate removed was conducted at different time intervals and tested at $\lambda_{max} = 220$ nm using UV spectrophotometer, the results are given in Fig. 16.7. From Fig. 16.7, it is observed that the percentage of nitrate removed (% efficiency) decreases as the initial concentration of nitrate solution increases for all the tested concentrations and becomes constant after 5 days contact time, where saturation achieved and the sorption process attained equilibrium. The adsorption rate after the second day of contact time could be explained by activation that occurred to chia seeds after adding citric acid and its effect on activating the surface charges of chia seeds, especially since after burning chia seeds, the components of the seeds are positively charged oxides, and this has a significant role in attracting the negative. Charged nitrate ions on the surface layer of charged chia seeds. Positive based on Van der Waals strength [2], Also, the surface area of chia seeds after converting them to activated charcoal has a role in trapping nitrate ions in the cavities of charred seeds. The maximum amount of nitrate removed at each initial concentration of nitrate in mg/g of the Chia seeds biosorbent at day 5.

The decrease in % nitrate removal by increasing concentration (Figs. 16.7 and 16.8) can be explained by the absorption of adequate amount of nitrate from the adsorbate (nitrate solution) according to the available number of adsorption pores or active sites within the fixed amount of Chia seed adsorbent dose. In other words, the substrate becomes saturated at a certain concentration of adsorbate [12]. The sorption also normally controlled by the diffusion process from the bulk to the surface which is related to nitrate concentration [3]. With the exception of 1000 and 1500 ppm initial concentrations, the increasing concentration gradient acts as increasing driving force to overcome all mass transfer resistances of the nitrate between the aqueous and solid phase, leading to an increasing equilibrium sorption until sorbent saturation is achieved. Similar trend has been reported for chromium and fluoride removal by geo-materials [4] and activated charcoal [12].

Conclusion

This study investigated the adsorption capacity of Chia seeds char as a potential adsorbent for the removal of nitrate from aqueous solutions using batch system. The experimental parameters were very important in order to understand the mechanism of adsorption of nitrate ions, such as pretreatment methods of biosorbent, the initial nitrate ion concentration, adsorbent dose, pH and particle size. The results showed that:

1. Chia char was an effective biosorbent for nitrate removal comparing with other reported adsorbents.
2. Nitrate in synthetic and real groundwater could be effectively reduced by the Chia char biosorbent process with a nitrate removal rate of about 93%.
3. The results showed that the removal efficiency of nitrate related to particle size and the best particle size was of 2 mm of raw Chia char.
4. Experimental data showed that nitrate adsorption increased by decreasing pH value of the solution and the optimum pH was at (5.8–6).
5. The results showed that equilibrium time for nitrate removal was after 5 days.
6. The removal efficiency increases with increasing biosorbent dose.
7. For recovery Chia char biosorbent, the results showed that the removal efficiency is 65% in case of washed samples while the removal efficiency is 78% in case of unwashed samples.
8. The results showed the efficiency removal of the use of Chai char in the process of treatment of nitrate ion in real samples was between (78–93%).

9. Through comparisons between chemical and thermal treatment of chia, it was found that heat treatment has a much better effect than chemical treatment of chia on nitrate removal, since thermal burning at a temperature of 650 °C gave better results than burning using chemical acids and bases.

Recommendations

Although the strategy used in this work has given very promising results, there are number of areas that need further investigation. These include:

1. Further testing Chia char biosorbent for other un-studied pollutants in water and groundwater should be conducted.
2. More analysis is recommended for the water quality after treatment with the Chia char biosorbent. Thus further researches on other pollutants using this technology will not only improve its efficiency, but also develop new modeling techniques.
3. More study on the economic feasibility of the Chia seed biosorbent preparation.
4. Further research investigation to promote a large-scale utilization of neglected natural resource for water treatment through filtration is needed.
5. Using natural media for water treatment applications are strongly recommended due to their local availability, an environmentally friendly and cost-effectiveness.
6. Investigation of using different forms of filter modifications by using another form of media toward the enhancing of the water treatment efficiency and speed up the removal process.
7. Further study and research should be carried out on the development of water filters to improve upon its performance in water treatment by the government, nongovernmental and private sector to make it available in most homes thereby alleviating the issue of inadequate supply of safe water in our local area.
8. A particular sorbent which shows high uptake of NO_3^- in the laboratory, may fail in field conditions. Thus, the selection of the appropriate technology/sorbent media can be tedious.
9. Besides these, some other issues, such as assessment of efficacy of sorbents for NO_3^- removal under multi-component pollutants, investigation of these materials with more real water samples under different concentration and TDS, and continuous flow studies should be conducted in detail.
10. Last but not the least; it would be worthwhile to investigate the reusability of the spent adsorbents as only a few studies are available in literature. More research is needed in the field of regeneration and finally for the environmentally safe disposal of NO_3^- laden adsorbents.

References

1. Afkhami, A.: Adsorption and electrosorption of nitrate and nitrite on high-area carbon cloth: an approach to purification of water and waste-water samples. Carbon 6(41), 1320–1322 (2003)
2. Al-Anber, M.A.: Thermodynamics approach in the adsorption of heavy metals. Thermodynamics-Interaction Studies-Solids, Liquids and Gases. InTech (2011)
3. Bhaumik, R., Mondal, N.K., Das, B., Roy, P., Pal, K.C., Das, C., Baneerjee, A.: Eggshell powder as an adsorbent for removal of fluoride from aqueous solution: equilibrium, kinetic and thermodynamic studies. J. Chem. 9(3), 1457–1480 (2012)
4. Bhatti, H.N., Nasir, A.W., Hanif, M.A.: Efficacy of *Daucus carota* L. waste biomass for the removal of chromium from aqueous solutions. Desalination 253(1–3), 78–87 (2010)
5. Grassi, M., Kaykioglu, G., Belgiorno, V., Lofrano, G.: Removal of emerging contaminants from water and wastewater by adsorption process emerging compounds removal from wastewater, pp. 15–37. Springer (2012)
6. Hagerty, P.A., Taylor, J.R.: Nitrate removal for on-lot sewage treatment systems: the POINTTM System (2012). http://www.taylorgeoservices.com/papers/point%20system
7. Jendia, A.H., Hamzah, S., Abuhabib, A.A., El-Ashgar, N.M.: Removal of nitrate from groundwater by eggshell biowaste. Water Supply 20(7), 2514–2529 (2020)
8. Kumar, P.S., Ramalingam, S., Senthamarai, C., Niranjanaa, M., Vijayalakshmi, P., Sivanesan, S.: Adsorption of dye from aqueous solution by cashew nut shell: studies on equilibrium isotherm, kinetics and thermodynamics of interactions. Desalination 261(1–2), 52–60 (2010)
9. Mae, K., Uno, A., Maki, T.: A new removal method of nitrates ion in wastewater by pH-swing reaction under hydrothermal condition. In: Asian Pacific Confederation of Chemical Engineering, Congress Program and Abstracts, pp. 445–445. The Society of Chemical Engineers, Japan (2004)
10. Martínez-Navarrete, C., Jiménez-Madrid, A., Sánchez-Navarro, I., Carrasco-Cantos, F., Moreno-Merino, L.: Conceptual framework for protecting groundwater quality. Int. J. Water Resour. Dev. 27(1), 227–243 (2011)
11. Michalski, R., Kurzyca, I.: Determination of nitrogen species (nitrate, nitrite and ammonia ions) in environmental samples by ion chromatography. Pol. J. Environ. Stud. 15(1) (2006)
12. Murugan, M., Subramanian, E.: Studies on defluoridation of water by Tamarind seed, an unconventional biosorbent. J. Water Health 4(4), 453–461 (2006)
13. N'goran, K.P.D.A., Diabaté, D., Yao, K.M., Gnonsoro, U.P., Kinimo, K.C., Trokourey, A.: Lead and cadmium removal from natural freshwater using mixed activated carbons from cashew and shea nut shells. Arab. J. Geosci. 11(17), 498 (2018)
14. Öztürk, N., Bektaş, T.E.: Nitrate removal from aqueous solution by adsorption onto various materials. J. Hazard. Mater. 112(1–2), 155–162 (2004)
15. Palestinian Water Authority, PWA: Evaluation of Groundwater-Part B-Water Quality in the Gaza Strip Municipal Wells. Water Resources Directorate (2013)
16. Shrimali, M., Singh, K.P.: New methods of nitrate removal from water. Environ. Pollut. 112(3), 351–359 (2001)
17. Torrentó, C., Cama, J., Urmeneta, J., Otero, N., Soler, A.: Denitrification of groundwater with pyrite and Thiobacillus denitrificans. Chem. Geol. 278(1–2), 80–91 (2010)
18. WHO, U., Mathers, C.: Global strategy for women's, children's and adolescents' health (2016–2030). Organization 2016(9) (2017)
19. Zyoud, A., Nassar, H.N., El-Hamouz, A., Hilal, H.S.: Solid olive waste in environmental cleanup: Enhanced nitrite ion removal by ZnCl_2-activated carbon. J. Environ. Manage. 152, 27–35 (2015)

Chapter 17
Smart Agriculture Research and Development for Small Island Developing States

Kavi Khedo and Avinash Mungur

Introduction

Small island developing states (SIDS) have several common characteristics that pose severe challenges to their development. SIDS differ considerably in their demographic, geographic and economic characteristics, however their common physical characteristics of small size and remote location make them extremely vulnerable to environmental and economic shocks. For example, SIDS have been hit the most by the COVID-19 pandemic. Moreover, SIDS are the first to suffer the consequences of climate change for which they bear little responsibility. SIDS are caught in a cruel paradox: they are collectively responsible for less than one percent of global carbon emissions, but they are suffering severely from the effects of climate change, to the extent that we could become uninhabitable. There is a pressing need to reduce vulnerabilities of SIDS and to build resilience to external shocks.

One of the major concerns for SIDS is food security. Recent happenings such as the COVID-19 pandemic and the war in Ukraine have exposed the excessive dependence of SIDS on food imports. The Food and Agriculture Organization (FOA) of the United Nations predict that the world will have to produce 60% more food globally, and 100% more in developing countries by 2050 to sustain the growing population of the Earth [2]. The food production industry is therefore becoming more important than ever. To meet this high demand, farmers in SIDS need to gradually turn to the use of new technologies to increase production capabilities while minimizing cost and preserving resources.

K. Khedo (✉) · A. Mungur
Faculty of Information, Communication and Digital Technologies, University of Mauritius, Moka, Mauritius
e-mail: k.khedo@uom.ac.mu

A. Mungur
e-mail: a.mungur@uom.ac.mu

© Centre for Science and Technology of the Non-aligned and Other Developing Countries 2023
K. Pakeerathan (ed.), *Smart Agriculture for Developing Nations*, Advanced Technologies and Societal Change, https://doi.org/10.1007/978-981-19-8738-0_17

Globally, there is a need to make food production efficient and sustainable by reducing both the use of resource-intensive inputs such as fertilizers, pesticides and freshwater, and negative outputs such as water pollution and soil loss [8]. Without improving yields, 70% increase in food would require over 34,000,000 km^2 of new agricultural land, an area larger than the entire continent of Africa [7]. Countries that produce large amounts of food such as India, China, Australia, Spain and the US are close to reaching their water resource limits [31].

Given the limited natural resources of SIDS, there is a need to use high technological farming techniques and technologies such as agricultural drones and sensors to monitor environmental data such as soil temperature, soil moisture, humidity, light, ambient temperature and carbon dioxide to facilitate precision agriculture. Precision agriculture has the potential to promote yield by 1.75%, reduce energy costs by an average of $10 per acre, and reduce water use for irrigation by 8% [17]. Climate changes such as heavy rainfall, more intense storms and an increase in atmospheric temperature can be disastrous for the agri-food production sector in SIDS. To mitigate and adapt to climate change to make food production sustainable, SIDS need to adopt smart agriculture practices and should embrace IoT with ICT-based decision support systems. In this chapter, the unique agricultural challenges faced by SIDS are exposed and emerging agricultural technologies that can help to overcome the challenges are identified and discussed.

SIDS Agricultural Challenges

The agricultural challenges for SIDS stem from their inherent characteristic of small landmass, remoteness and large marine resources that are highly exposed to the warming of the oceans. These island nations have a very limited agri-food production sector and are mostly dependent on fisheries [26]. They face several agricultural limitations resulting in a heavy reliance on imports. Their heavy dependence on food imports exposes them to issues of food prices volatility at the international level and food insecurity. SIDS are also facing sea-levels rise and reduction of freshwater sources that are aggravating the agricultural challenges and food security issues. Moreover, over the years many island states have directed resources away from the agricultural sector to develop the tourism sector [25]. Consequently, due to the heavy dependence on tourism, the economic situation of these island states has worsened dramatically with the COVID-19 pandemic.

Climate Change and Natural Disasters

Climate change and associated sea level rise are negatively impacting agriculture in SIDS and a recent assessment indicates that this is set to worsen. Indeed, climate change is having severe adverse impacts on agricultural production in SIDS which are seriously challenging food security in these island states.

There are several ways in which climate change is contributing to food insecurity in SIDS namely drastically reducing food production and increasing food prices. Since climate change mitigation activities are increasing energy prices, food production is becoming more expensive. Due to extreme weather conditions such as frequent drought, water resources required for food production are becoming scarcer in SIDS [6]. Once fertile land in SIDS is increasingly becoming climatically unsuitable for agricultural activities. Moreover, the high variations in weather conditions associated with climate change are causing drastic reductions in agricultural productivity and significant increase in food prices. SIDS have always been exposed to natural disasters such as cyclones and storms for centuries, however climate change is aggravating their intensity which is even causing an existential threat to some island states. Flash floods and prolonged droughts are very frequent in SIDS, and this is severely impacting on the agricultural sector.

Food Insecurity and Land Use

The concurrent COVID-19 and War in Ukraine crisis has exposed the unreasonable dependence of SIDS on food imports. In recent years, many SIDS have significantly reduced traditional food production including agriculture and had increased investments in tourism and urban development. Consequently, in the period of crises, SIDS are more vulnerable to fluctuations in food prices at the international level. This situation is urging Governments of SIDS reconsider their strategies with respect to development and redirect resources to local food production.

With the limited land resources in SIDS, efficient land use is essential to tackle the food security problem [14]. There is a high demand for land for other economic sectors such as the tourism industry and foreign investors among others, therefore SIDS will have to implement appropriate land use policies. A holistic approach is required to address the problem of adequate high quality and nutritious food supply in SIDS. For example, strategies to encourage local communities to explore organic methods for the production of high-quality food should be put in place. SIDS has the potential to become leaders in new forms of agriculture such as hydroponics that use less land. At the international level, more appropriate policies and agricultural subsidies should be developed in support of SIDS food security.

Smart Agricultural Research and Innovation for SIDS

Recently, there has been significant development in agricultural technologies that may help SIDS to build resilience to climate change, increase food production and overcome the challenges associated with limited natural resources. Innovative agricultural technologies and practices including precision agriculture, climate smart agriculture, Internet of things, artificial intelligence, data analytics and drone technologies have great potential to alleviate most of the unique agricultural challenges of SIDS.

Indeed, smart agricultural research and innovation have the potential to increase production, whilst reducing environmental, energy and the climate footprint of food production in SIDS. More investments should be directed toward smart agricultural research and innovation so as to develop agricultural technologies that will help SIDS to overcome their unique agricultural challenges. The technological revolution in farming is led by an ever better understanding of the biology of plants and advances in robotics and sensing technologies. Through intelligent digital devices with highly advanced image and sensor technology, such as robots and drones, monitoring and maintenance of soil quality can be improved, helping to eliminate pests and diseases while further reducing the need for agrichemicals.

The rapid urbanization in SIDS is placing enormous demands on urban food supply systems. Urban agriculture can solve such problems by converting urban wastes into productive resources, for example, the conversion of organic waste to compost [30]. Technologies such as hydroponics, green house, glass house, aquaponics and horticulture are very interesting for urban environments in SIDS. The different urban agricultural setups will require advanced automation and control which again illustrate the need for smart agricultural research and innovation. Additionally, agricultural research is required for organic or bio-farming that involves growing crops without pesticides for a healthier population. It is widely known that excessive use of pesticides on crops increases risk not only to human health but also to the environment.

For example, agricultural research and innovation in sensing technologies and internet of things (IoT) have great potential to transform the agri-food production industry in island states that have the challenge to overcome water shortages, limited agricultural lands, and difficulty to manage costs while meeting the needs of the growing population. Farms can use IoT sensors to remotely monitor changing environmental conditions such as soil moisture, pH levels, temperature, etc. while attempting to automate processes such as irrigation. Kim et al. pointed out the need of convergence toward technology to achieve precision agriculture based on the dynamic environmental parameters to make agriculture sustainable for the future [12]. Artificial intelligence can also be used to provide new insights and facilitate decision making enabling farmers to easily visualize data and take action on insights and recommendations. There are also a lot of IoT agricultural systems developed both in the research as well as commercial domains that may be very useful to SIDS, such as ThingWorx, OnFarm systems that relied entirely on ThingWorx'sIoT platform, CropX and FarmBot.

ThingWorxIoT platform collects and manages big data from sensors on the cloud, offers innovative IoT applications, leverages analytics on big data to provide recommendations to help farmers in better decision making [24]. Kaa, an open-source IoT platform is a middleware technology that streamlines the development of smart farming systems such as field, crop, agricultural equipment and climate monitoring, prediction analytics for crops while providing smart logistics [11]. Both CropX offers an adaptive irrigation system that automatically optimizes irrigation based on the needs of the soil and the crops by measuring environmental variables and making use of predictive algorithms [5].

Research and innovation are also required to make agricultural systems more intelligent and help farmers in SIDS decision making to increase food productivity [18]. There is very limited use of computational intelligence in existing agricultural systems. Computational intelligence (CI) is the discipline of replicating human intelligence on computers and it can provide the missing knowledge-based layer on the various existing agricultural systems. Therefore, there is an urgent need to exploit techniques and algorithms of computational intelligence, machine learning and/or artificial intelligence in those systems, such as case-based reasoning, rule-based reasoning, artificial neural networks, fuzzy models and swarm intelligence. With the adoption of these techniques and algorithms, the real time processing of sensor data and the real time reasoning in the IoT-enabled precision agriculture systems can definitely shape the future of agriculture in SIDS and help farmers in meeting the increasing food demands [17].

Another area that requires extensive research and innovation is climate smart agriculture (CSA). CSA is a holistic approach of managing the agricultural landscapes including fisheries, livestock and cropland while addressing the challenges related to accelerating climate change and food security [13]. CSA has the potential to help SIDS in achieving three very important outcomes including increasing food productivity and improving nutrition security; enhancing resilience against climate related risks and vulnerability to diseases, pests, drought, and reducing emissions by pursuing lower emissions for each kilo of food produced. Indeed, CSA can help SIDS to systematically consider the tradeoffs that exist between productivity, adaptation and mitigation.

Emerging Agricultural Technologies

This section discusses the emerging technologies which may support the agricultural sector in the Small Island Developing States. The emerging technology has greater potential to increase food production and allow SIDS to build resilience to environmental and economic shocks. The emerging technologies such as drone, artificial intelligence and IoT will be elaborated and their adoptions to SIDS are discussed.

Drones

Drones, also known as unmanned aerial vehicles (UAVs) are increasingly being used in the agricultural sector due to their relatively low cost and ability to efficiently take pictures and collect data on crops and animals. Drones come in various sizes and

Fig. 17.1 Multispectral
health of leaves
(©agribox.com, Sylvester
[22])

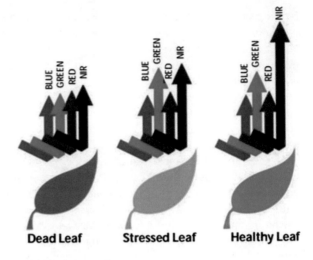

weights. Depending on the application, the appropriate drone is used. However, all drones come with a digital camera in order to capture images or videos. Applications of drones in SIDS are as follows (and not limited to):

- Monitoring the health of crop
- Surveying soil and land conditions
- Pest and disease control.

Crop Health Monitoring

To monitor the health of crops, a drone with a high resolution camera embedded with sensors to capture both visible light (Red, Green and Blue) and near-infrared (NIR) light are used. Usually the sensors are termed as multispectral or hyper spectral which can detect light in discrete chunks and are used for monitoring specific [3]. When a drone flies over vegetation, the sensors will be capturing reflectance information of plants (which is the ultraviolet radiation (NIR) generated by the plants). Those data which have been collected could be used to classify images into leaves, soil, plants and they would provide valuable information on the physiology and form of the crop. In addition, one significant information which is captured is the chlorophyll values from plants' leaves (through the strong reflectance values (NIR) generated). When the reflectance value is low, this indicates that the plant is dehydrated or dead. Otherwise, a high reflectance value indicates strong growth rate. Figure 17.1 shows an example how we can differentiate among the health states of the leaves.

Moreover, through the imagery data, it will be easier to detect areas where growth has been delayed due to underlying soil conditions or environmental conditions. Quick actions can be taken such as re-seeding the area before the actual harvest starts. Also the real time monitoring of the various vegetative stages will contribute to a high yield, decrease the use of fertilizers and allow farmers to experiment with

Fig. 17.2 Tea plant mapping
(©TCS, Sylvester [22])

other types of crops to diversify their production. Furthermore, depending on the type of crop, a farmer can at an early stage estimate his yield and overall productivity by counting the number of flowers or fruits at a specific stage. This will enable the farmers to adapt their farming strategies if ever they find that they are not reaching their required productivity such as adapting their irrigation or fertilizers treatment.

Thus, the use of drones will enable a dynamic crop health monitoring which would have a direct economic impact on the farmers and their livelihood. Instead of treating a crop at its last stage of its health, it is better to intervene as soon as a problem is detected. This will save cost and increase yield production.

Surveying Soil and Land Conditions

Drones have extensively been used to make a rapid assessment of the soil and land condition after the occurrence of cyclones, floods and landslides (which are frequently being faced by SIDS). Through their imaging and sensing capabilities, drones can quickly detect soil erosion, soil PH and nutrients deficiencies [20]. By detecting those latter issues at an early stage, remedial action can be taken. Thus, this will avoid expensive and time-consuming actions if they were to be discovered at a later stage [22]. The impact of an early detection will significantly improve the crop yield. In addition, using the imagery data, various crop maps can be produced which will enable farmers to better plan their seeding processes. They will know the appropriate area of land which will be beneficial to start the seeding process and which area of land to avoid as shown in Fig. 17.2. Furthermore, through the frequent monitoring of the land, the surrounding infrastructure can be monitored such as the fencing or other farm equipment (Bendig 2012).

Pest and Disease Control

Drones have become an important tool in fighting against insect infestation, fungal diseases and bacterial invasion. Through the use of the multispectral sensor

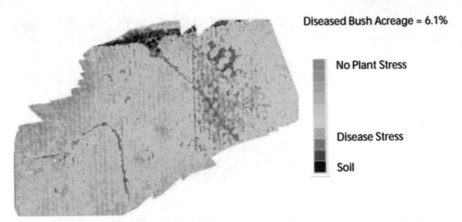

Fig. 17.3 Pest incidence (©TCS, Sylvester [22])

(embedded in the drones) which regularly scans the vegetation and capturing the reflectance values, it is easy to distinguish between healthy and diseased leaves. The early detection of unhealthy leaves will immediately trigger remedial actions and also guards against the spreading of a potential viral infection which can destroy an entire harvest. Subsequently the early detection of any disease will have an environmental impact since only a minimum amount of pesticides will be required. The fear of an excessive amount of pesticides to be washed away will also be decreased. This will mitigate water pollution caused by the over application of pesticides. In the same vein, the data provided by the drone will enable the correct identification of the disease and the correct treatment to be applied (the correct pesticide to be used). This will avoid any wrong diagnosis to be made and prevent any damages caused through the wrong application of pesticides. Distinguishing between a disease and a weed invasion is primordial since this will have an effect in the treatment and in the correct selection of herbicides, insecticides and pesticides. Identifying the correct treatment is important in order to avoid any environmental damage [21]. Figure 17.3 shows a pest incidence map which illustrates the area under which there might be a potential disease outbreak.

Regulatory Framework for Drones in SIDS

Operating a drone requires special clearance from a regulatory body. Usually it is the Civil Aviation body which delivers a clearance to fly the drone. Depending on the weight of the drone, the drone operator should ensure that all safety precautions are observed so as to minimize public harm [1]. The drone operator should adhere to the license obtained from the regulator. For example, drones should be flown at a certain time of the day and at a certain altitude. It should be kept at a certain distance from a public place and from an airport. It should not be used to cause any nuisance or trespass any perimeter.

Internet of Things

Internet of Things (IoT) is a disruptive technology which is changing the way agriculture is being conducted. IoT is a paradigm shift toward a sustainable and ecological way of performing agriculture [18]. In essence IoT is an ecosystem which is composed of sensors, communication protocol, cloud platform and analytic and end user technologies. Such an IoT ecosystem enables the real time monitoring of plants and animals in order to detect any abnormalities at an early stage and to increase productivity as well as automating the agricultural process. Sensors are intensively being used to collect vital information such as the plant condition, soil moisture level, humidity level, temperature of the surrounding and the presence of chemicals and gasses. Those data are captured and transmitted to a cloud storage system on the Internet through different communication protocols [11]. Those data are analyzed using AI in order to provide informed decisions to the farmers. The importance of IoT as a whole helps to provide useful and real time information which would assist farmers in their day to day activities.

- The application of IoT in agriculture and farming is wide and diverse:
- The data collected and analyzed in an IoT system are used to automate the irrigation process [23] whereby water is used in a sustainable manner. Thus water is provided to the right amount which helps to reduce cost, reduces water waste and prevents waterlogging.
- IoT is used in greenhouses to manage the temperature and moisture automatically. This ensure the optimal condition for plant growth
- IoT is applied in vineyards and horticulture to monitor environmental conditions
- IoT is widely used in shed (livestock shed) to monitor the enclosure in order to

    - To detect stress and improve animal welfare.
    - To timely detect a disease outbreak and to provide early veterinary treatment.
    - To improve the reproduction rate. Through the use of sensors, we can know exactly when a cow is in heat and insemination can take place. The next heat cycle can also be predicted.
    - Optimize the feeding of the livestock (prevent under or over-feeding).
    - Geo-localization of livestock is made possible.

- IoT is becoming an important tool in order to monitor livestock from birth to slaughter so as to provide traceability and to be compliant with food and safety regulation (it will be possible to know whether there has been a breach in the cold chain thereby ensuring the safety and quality of the product).

Adoption of IoT in SIDS

The enabling factor which would allow the implementation of the IoT ecosystem relies on the ICT infrastructure put in place in the country. The ICT infrastructure should at least support wireless technologies in the form of WIFI and 3G. Mobile

Broadband is also an important ingredient in the IoT ecosystem. In nearly 21 SIDS countries, the mobile broadband coverage has improved significantly since 2014 reaching around 85% of the population in 2018 [10]. Such broadband coverage will encourage investment in the development of IoT systems. Furthermore, the importance of the SIDS to be connected to the undersea fiber-optic has been recognized. By 2019, 32 countries out of 39 SIDS are connected to at least one submarine fiber-optic cable [10]. Being connected to the cable will only accelerate the adoption of IoT. An improved ICT infrastructure will enable access to the Internet and will improve the digital literacy of the country. This will create a conducive environment to boost IoT-enabled systems.

Regulatory Framework for IoT in SIDS

Even though many SIDS have good Internet connectivity, the proliferation of IoT systems in SIDS is relatively quiet slow. One of the main stumbling blocks is on the Data Ownership [1]. Since sensors collect a wide variety of data, the question of who owns the data becomes a significant issue because there are different stakeholders involved such as the farmers, the software and hardware providers. In addition, the retention and access to the raw data can become a cause of concern. For example, a bank or insurance can use the raw data to predict the productivity of a farmer before sanctioning a loan or insurance. In another example, the health data of livestock could be used to adversely affect farmers. To dissipate the latter concerns and to accelerate the adoption of IoT, a regulatory framework should be put in place which will take into account the Data Protection and Cyber Laws surrounding the use of data in an IoT ecosystem. This will enable users to have trust in the system. However, it must be said that many SIDS countries have strong laws regarding the use of data which can be amended to cater for IoT data.

Artificial Intelligence

Artificial intelligence is slowly transforming the agricultural sector as a whole. AI covers a wide range of technologies such as machine learning, neural network, deep learning and blockchain. The main idea behind AI is the ability to perform tasks without human intervention. AI is especially employed to analyze data and to provide informed decisions based on the data and offers prediction capabilities. For example, all the data that has been captured by drones and sensors in an IoT system are analyzed using AI. Those analysis are performed quickly and trends are derived to assist in the decision making process.

AI has improved crop production and enhanced the real time monitoring. It has contributed positivity in the food industry. The use of AI in agriculture has opened up a new branch in agriculture namely Precision agriculture [29]. However, AI is a supporting technology which is used along with other technologies and hardware to automate the agricultural process. For example, AI is intensively used in a smart irrigation system. Sensors which are capturing soil moisture data are transmitted to an analytic platform where AI is applied. The output of the analysis can determine whether to start or stop an irrigation process automatically and without the intervention of the farmer. Similar process happens when drones capture real time images and are transmitted for processing. Analyzing those images without AI will be impossible. This is because various techniques should be applied to the image such as image recognition and perception in order to detect or identify a color on a leaf, pest or disease.

Another application of AI is in chatbots for farmers. Chatbots are virtual assistants which automate the interaction with end users using the natural language of the user. Figure 17.4 shows a chatbot interaction. Different chatbots have been created to assist farmers during their cultivation process. For example, authorities provide farmers with chatbots which they can use to query on the properties of a plant or animal and they will be provided with advice and recommendation.

In addition, AI is used to control robots which perform various agricultural operations in order to increase efficiency when compared to traditional machineries. AI

Fig. 17.4 A chatbot conversation [28]

Fig. 17.5 Blockchain technology [16]

enabled robots perform different tasks efficiently and autonomously such as removal of weeds, sowing, irrigation and surveillance.

Blockchain is another emerging technology which is used in the agricultural industry in order to provide a traceability trail of the food from its creation until it comes to our plate. Customers are very careful to avoid consuming food which has been produced in an unethical way and without following proper agricultural standards. In order to follow the whole supply chain from the producers, through the suppliers until it reaches the client, blockchain can be used. It will provide a traceability aspect which will ensure quality as shown in Fig. 17.5.

The contribution of AI to agriculture is enormous because it has an impact on the harvesting process, crop yield and productivity. It has decreased the intervention of humans enabling the farmer to concentrate on other aspects of farming. This has reduced the burden of farmers whereby labor is an issue.

Adoption of AI in SIDS

AI is an enabling technology. Since it is used along with other software and hardware, it will have similar issues such as the data ownership. This issue can be tackled through appropriate Data protection law. However the major concern is on the autonomous decision making aspect. Determining the legal liability in cases where AI has taken a decision which has an adverse effect on someone can be complicated [1]. The latter issue is not so evident in the agricultural sector. To better comprehend the need of AI in a SIDS country, it will be recommended to set-up an AI council [9]. Mauritius has set-up an AI council in order to look into various aspects of AI and its applications. Based on the local needs, the council provides numerous advice and recommendations on the use of AI.

Case of the University of Mauritius

Mauritius along with Cabo Verde, Guinea-Bissau, Sao Tome and Principe and Seychelles are the five SIDS found in the African region. Mauritius is situated in the Indian Ocean and has an area of 2040 square kilometers and a population of 1.2 million inhabitants. The main drivers of the Mauritian economy are the Tourism sector, Manufacturing sector, Agricultural sector, Financial sector, Information and Communication Technology sector among others. Agriculture in Mauritius is predominantly dominated by sugar cane fields. Nevertheless, the agricultural sector has been diversifying to include other crops [15]. In the quest to improve the agricultural sector, the University of Mauritius, which is one of the main research institutions in Mauritius, has massively contributed in terms of research and development as well as introducing innovative concepts to this sector. Thus this section will provide an overview of the contribution of the University of Mauritius in promoting and adopting ICT-based technologies in the agricultural sector. The University of Mauritius recognized the fact that all SIDS face similar challenges as Mauritius. Thus it is important to boost the adoption of technologies in the agri-sector.

Agri-digital Skills

The University of Mauritius is a leading tertiary institution which is engaged in promoting research and intellectual talent to support the needs of the country. In Mauritius, agriculture contributes around 3.41% of the GDP and 6% of the population works in the agricultural sector [4]. In order to support the agricultural sector and to promote the use of technology in this sector, the Faculty of Agriculture (FoA) in joint collaboration with the Faculty of Information Communication and Digital Technologies (FoICDT) are mounting interdisciplinary academic programs which include modules such covers the use of emerging technologies [27]. In addition, various PhD research is conducted in the field of agriculture with the support of ICT. Training and short courses are also offered in digital agri-skills and digital entrepreneurship [4] in order to foster the agri-e-commerce culture.

Moreover at the FoICDT, numerous prototypes are developed to tackle agricultural issues and to demonstrate their viability and applicability. For example, prototypes such as smart farming systems, smart irrigation systems, smart greenhouse systems, smart livestock monitoring systems and smart pest control systems are developed to tackle specific agricultural issues. Other prototypes are also developed using the above mentioned emerging technologies. The overall aim of both Faculties is to equip the student with the digital knowledge and digital skills to accelerate the adoption of technologies in the agricultural sector.

AgriTECH Park

In March 2020, the University of Mauritius launched the AgriTECH Park, in order to boost collaboration between Academia, the private sector, SMEs, Start-ups and Incubators, and supportive of entrepreneurs in the development of innovative products and services for the Agri-Food sector in Mauritius and in the region [4, 27]. The focus of the AgriTECH Park will be around smart agriculture, biotechnology and food security.

The AgriTech Park will host an Agri-Processing Incubator which will assist in turning ideas and prototypes into commercial products. It will also be a place where students will gain valuable hand-on experience on real agricultural problems and foster a new breed of agri-technopreneur. A digital culture of start-up around the agricultural sector will be promoted targeting smart farming, smart hydroponic, smart aquaponics and smart livestock businesses.

The AgriTech Park will also provide an environment to conduct research through the set-up of innovative labs where research will be carried out with local, regional and international collaborators. The research output will be shared among all SIDS.

Multidisciplinary Research

The University of Mauritius is engaged in various multidisciplinary and transdisciplinary research in agriculture. One such research is the Development Smart Innovation through Research in Agriculture (DeSIRA) project which is funded by the European Union in the field of smart agriculture. The aim of the DeSIRA project is to research and develop smart and innovative solutions in the field of agriculture to adapt to climate change in Mauritius. Through the DeSIRA project, it is expected to conduct research and produce scientific knowledge and prototype to enhance the innovation capacities and techniques which stakeholders (farmers, private and public sectors, universities among others) can use to reproduce. This will encourage the stakeholders to adopt such techniques in order to increase their productivity [4, 19, 27].

The University of Mauritius is extensively involved in searching new avenues in order to promote the use of the technologies in agriculture by collaborating with local, regional and international shareholders. For example, the University of Mauritius has signed several MoU with local ICT companies in order to develop prototypes and research in the agricultural sector. Those types of collaboration show that the University of Mauritius is finding other means of bringing technology to the sector. Thus the University of Mauritius is enlarging its approach to transdisciplinary research. The output of those research will contribute to knowledge space among the SIDS.

Conclusion

Almost all of the SIDS share similar characteristics and are vulnerable to climate change. The latter has a direct impact on the environment and on the ability for producing food locally. With the rising inflation around the world and the impact of the war, there is a pressing need for SIDS to decrease their dependency on importing food and to start producing them locally. Most SIDS have the capacity to become self-sufficient by improving their agricultural sector. In order to enhance the agricultural sector in SIDS, the use of smart innovative agricultural technologies and conducting research is essential. As a result, this chapter provides an insight of the main agricultural challenges encountered by SIDS and provides an overview of the main emerging technologies such as drones, IoT and AI in agriculture. The opportunities and challenges of the emerging technologies are highlighted with respect to their adoption to SIDS. The importance of conducting innovative multidisciplinary research and building digital skills set in the field of agriculture are also discussed through the University of Mauritius case study. Undoubtedly, the opportunities provided by the use of smart innovative agricultural technologies and research will contribute largely to a sustainable agri-food sector in SIDS.

References

1. AgriFutures Australia: Emerging technologies in agriculture: regulatory and other challenges (2018)
2. Alexandratos, N., Bruinsma, J.: World agriculture towards 2030/2050: the 2012 revision. FAO: ESA Working paper, 12-03 (4) (2012)
3. Bendig, J., Bolten, A., Bareth, G.: Introducing a low-cost mini-UAV for thermal-and multispectral-imaging. Int. Arch. Photogramm. Remote Sens. Spat. Inf. Sci. **39**, 345–349 (2012)
4. CCARDESA: Digital Agriculture County Study—Mauritius (2022)
5. CROPX. Cloud-based technologies to boost crops yield while saving water and energy (2022). http://cropx.com. Accessed 12 June 2022
6. Crumpler, K., Bernoux, M.: Climate change adaptation in the agriculture and land use sectors: a review of nationally determined contributions (NDCs) in Pacific Small Island developing states (SIDS). Managing Climate Change Adaptation in the Pacific Region, pp. 1–25 (2020)
7. FAO (UN Food And Agriculture Organization): Sustainable Food and Agriculture (2014). http://www.fao.org. Accessed 18 June 2022
8. Fisher, J.: Global Agriculture Trends: Are We Actually Using Less Land? (2014) http://blog.nature.org/science/2014/06/18/global-agriculture-land-sustainability-deforestation-foodsecurity. Accessed 18 June 2022
9. Georges, C.-T.-K., Acharuz, A., Moorghen, Y., Jhurry, D., Cartier, C., Suddhoo, A., Guibert, F., Mohee, K., Meetoo, A.: Mauritius Artificial Intelligence Strategy. Mauritius (2018). https://mitci.govmu.org/Documents/Strategies/Mauritius%20AI%20Strategy.pdf. Accessed on 29 July 2022
10. ITU.: Small Island developing states (SIDS) and ICTs Mid-term review of the Samoa Pathway (2019)

11. Kaa: Kaa multi-purpose platform for the Internet of Things (2022). http://www.kaaproject.org/agriculture. Accessed 15 June 2022
12. Kim, Y., Bae, P., Han, J., Ko, Y.B.: Data aggregation in precision agriculture for low-power and lossy networks. In: Communications, Computers and Signal Processing (PACRIM) IEEE Pacific Rim Conference, 24–26 August 2015 University of Victoria, Victoria, B.C., Canada, pp. 438–443 (2015)
13. Lipper, L., Thornton, P., Campbell, B., et al.: Climate-smart agriculture for food security. Nat. Clim. Change **4**, 1068–1072 (2014). https://doi.org/10.1038/nclimate2437
14. Lowitt, K., Ville, A.S., Lewis, P., Hickey, G.M.: Environmental change and food security: the special case of Small Island developing states. Reg. Environ. Change **15**(7), 1293–1298 (2015)
15. MCA (Mauritius Chamber of Agriculture) (2022). https://chamber-of-agriculture.mu/agriculture-in-mauritius/. Accessed 28 July 2022
16. Menon, S.: Blockchain Platform Will Revolutionize Agri-Food Supply Chains (2018). https://www.sathguru.com/news/2018/02/15/blockchain-platform-will-revolutionize-agri-food-supply-chains/. Accessed 28 July 2022
17. Meola, A.: Why IoT, Big Data & Smart Farming are the Future of Agriculture (2021). https://www.businessinsider.com/smart-farming-iot-agriculture?r=US&IR=T. Accessed 10 June 2022
18. Misra, N.N., Dixit, Y., Al-Mallahi, A., Bhullar, M.S., Upadhyay, R., Martynenko, A.: IoT, big data and artificial intelligence in agriculture and food industry. IEEE Internet of Things J. (2020)
19. One Planet Summit: DeSIRA (Development Smart Innovation through Research in Agriculture) (2022). https://www.oneplanetsummit.fr/en/coalitions-82/desira-development-smart-innovation-through-research-agriculture-206#:~:text=DeSIRA%20(Development%20Smart%20Innovation%20through%20Research%20in%20Agriculture)%20is%20the,the%20first%20One%20Planet%20Summit. Accessed 28 July 2022
20. Puri, V., Nayyar, A., Raja, L.: Agriculture drones: a modern breakthrough in precision agriculture. J. Stat. Manag. Syst. **20**(4), 507–518 (2017)
21. Rasmussen, J., Nielsen, J., Streibig, J.C., Jensen, J.E., Pedersen, K.S., Olsen, S.I.: Pre-harvest weed mapping of Cirsiumarvense in wheat and barley with off-the-shelf UAVs. Precision Agric. **20**(5), 983–999 (2019)
22. Sylvester, G.: E-Agriculture in Action: Drones for Agriculture. FAO and ITU, Bangkok (2018)
23. Talaviya, T., Shah, D., Patel, N., Yagnik, H., Shah, M.: Implementation of artificial intelligence in agriculture for optimisation of irrigation and application of pesticides and herbicides. Artif. Intell. Agric. **4**, 58–73 (2020)
24. Thingworx: Industrial IoT (2022). http://www.thingworx.com/ecosystem/markets/smart-connected-systems/smart-agriculture. Accessed 18 June 2022
25. United Nations Environment Programme (UNEP): Emerging Issues for Small Island Developing States (2014). https://sustainabledevelopment.un.org/content/documents/2173emerging%20issues%20of%20sids.pdf. Accessed 15 June 2022
26. United Nations: SIDS GBN—Thematic Areas—Sustainable Agriculture (2021). https://www.un.org/ohrlls/sids-gbn-thematic-areas-sustainable-agriculture. Accessed 20 June 2022
27. UoM (University of Mauritius): Launching of the UoMAgriTECH Park. Press Release (2020). https://docs.google.com/viewer?a=v&pid=sites&srcid=dW9tLmFjLm11fHZjLWJsb2d8Z3g6ZmQzYjM4MTU0MzA5MjRi. Accessed 28 July 2022
28. Vijayalakshmi, J., PandiMeena, K.: Agriculture TalkBot using AI. Int. J. Recent Technol. Eng. **8**(2S5), 186–190 (2019)
29. Wang, Y., Huang, L., Wu, J., Xu, H.: Wireless sensor networks for intensive irrigated agriculture. In: 2007 4th IEEE Consumer Communications and Networking Conference, pp. 197–201. IEEE (2007)

30. Weidner, T., Yang, A.: The potential of urban agriculture in combination with organic waste valorization: assessment of resource flows and emissions for two European cities. J. Clean. Prod. **244**, 118490 (2020)
31. World Wildlife Fund: Overview of water scarcity and causes (2017). https://www.worldwild life.org/threats/water-scarcity. Accessed 20 June 2022

Chapter 18
Smart Agriculture: Special Challenges and Strategies for Island States

Kandiah Pakeerathan

Introduction

Since the 1960s, in order to increase food production, the green revolution introduced high-yielding varieties that are now being cultivated intensively. But the ever-increasing human population requires increase in food production at the rate of 2% every year. Therefore, by 2050, we need to produce 70% more food from the limited land to feed the forecasted 9.1 billion people within the context of scarce natural resources and unexpected climate change. Overexploitation of existing natural resources in agriculture has deteriorated the physical and chemical quality of the soil, beneficial biota of the soil, water and environmental pollution, and carbon footprint. The environmentalists who are working zealously to create awareness to the general public on the importance of conservation of depleting precious natural resources suggest the sustainable agriculture or precision farming as the solution to the problem. The advances in information and communication technology opened a new era to modernize the agriculture sector as smart agriculture which integrates a variety of modern technologies, devices, protocols, and computational paradigms to improve agricultural processes [1].

Currently using smart technologies in agriculture, such as agriculture cyber-physical system (A-CPS), Internet-of-Agro-Things (IoAT), solar energy, sensor node, automatic crop disease prediction, machine learning, and convolutional neural network (CNN), are using microsensors and chips to decide depth and width of field ploughing, to detect soil nutrition availability and automatic application of required nutrition, to measure the soil and plant water availability and requirement, and to decide the irrigation intervals. Artificial intelligence (AI) is coupled with unmanned aerial vehicles (UAVs) and robots for fertilizer application, pest monitoring, disease

K. Pakeerathan (✉)
Department of Agricultural Biology, Faculty of Agriculture, University of Jaffna, Ariviyal Nagar, Kilinochchi 44000, Sri Lanka
e-mail: pakeerathank@univ.jfn.ac.lk

© Centre for Science and Technology of the Non-aligned and Other Developing Countries 2023
K. Pakeerathan (ed.), *Smart Agriculture for Developing Nations*, Advanced Technologies and Societal Change, https://doi.org/10.1007/978-981-19-8738-0_18

scouting/field phenotyping, pesticide application, harvesting, processing, and packaging. Chapters compiled in different books have explained at least one of the tools, devices, and their working principles, and the success stories of each application in different scenarios [2].

Advantages of Smart Agriculture

The advantages of smart agriculture include:

- Monitor farm/crop health in real time and identify crop diseases before they cause damage
- Conserve water with accurate insights into soil conditions and weather predictions
- Automate various farm operations like planting, harvesting, and spraying to reduce resource consumption and overall costs
- Evaluate precisely the actual production rates
- Reduce environmental footprint and increase carbon sequestration
- Improve livestock monitoring and management
- Increase sustainable productivity
- Strengthen farmers' resilience
- Reduce agriculture's greenhouse gas emissions
- Reduce the labor usage in farm operations.

Challenges of Smart Agriculture: Special Reference to Island States

Every innovation has its own pros and cons. Experts in the relevant fields are tirelessly working to make maximum benefits out of it. In smart agriculture too, challenges are common, and as far as small islands are concerned, the following special challenges need to be considered critically and handled strategically.

Mindset and Cultural Barriers

The majority of the farmlands in small island states are less than a hectare, and commercially cultivated lands are lease lands. A significant fraction of the farming community living in rural areas are more than 40-year-old males and are still believing and following traditional cultivation practices and approaches. Religious beliefs and patriarchal attitudes often restrict women's role in agriculture and resource access. Government support and incentive programs are often unsuccessful due to cultural and social barriers and land tenure [3, 4]. These factors significantly hinder the interest of the farmers to change their minds toward smart agriculture. Therefore, implementation of the smart agriculture techniques is a challenging task unless farmers' mindsets change.

Lack of Awareness on Smart Agriculture

Rapid industrialization and anthropogenic activities have caused tremendous change in the climate. In agriculture, increased flooding, extreme heat, drought, new pests, and disease attacks are posing economic loss. To hop-up from the extreme climate change, smart and precise farming is the ideal solutions. But the majority of smallholder farmers and marginal farmers in small island nations lack awareness on the benefits of smart farming as well as the application of digital agriculture for commercialization.

Scientific and Technical Knowledge of the Farmers

Smart agriculture integrates the modern cultivation aspects with technology and is also a connection between agricultural informatics, agriculture and entrepreneurship, agricultural facilities, technology diffusion, and information provided through the Internet and other relevant information communication [5]. In small island states like Sri Lanka, farmers have low literacy rate and scientific knowledge; therefore, they are not equipped to digital agriculture. Moreover, if a farmer wants to run a smart farm, he or she should have substantial basic knowledge on how sensor's (smart tools) reading data have to be monitored or interpreted properly to make necessary changes according to the crop requirement. The majority of the farmers are not showing a willingness to learn the technical aspects or are not ready to hire a technical person considering higher wages and the cost. Sorting out the operational issues of smart devices and tools, like proper maintenance, management, safety, and performance of smart devices, is also challenges; therefore, farmers are reluctant to adopt digital agriculture. The implementation cost of smart technology in the real field scenario is another deadlock to farmers [6].

Cost

Smart agriculture positively influences farm profit and sustainability in long run [7]. When considering the initial cost for the establishment of a smart farm, it will not be possible for all the farmers in developing countries unless sufficient financial assistance and technical support are given for a certain time period, at least until they get the first return. In small island states like Sri Lanka, Maldives, and Mauritius, financial assistance and research and development (R&D) are limited. Both the government and private sector involvements in R&D in technology should be strengthened by collaborating with more foreign institutions as well as with local universities. The country's marginal farmers may not be able to use these sophisticated technologies as yet; hence, agriculture companies need to implement these smart technologies at the field level through project like farmer participation in technology usage or signing of MoU for contract farming systems (https://www.ft.lk/Opinion-and-Issues/Smart-digital-farming-in-agriculture-Status-and-prospects-for-Sri-Lanka/14-691403).

In Sri Lanka, the Department of Agriculture (DOA) under the Ministry of Agriculture has taken initiatives to strengthen climate-smart agriculture and promote digital agriculture through agriculture modernization projects funded by the World Bank. A variety of ICT projects have now introduced several e-agriculture plans to overcome the challenges mentioned above (Sri Lanka E-agriculture Strategy, 2016), for example, the official website of the Department of Agriculture (www.doa.gov.lk); Wikigoviya website (www.goviya.lk); Krushilanka agriculture portal (www.krushilanka.gov.lk); Rice Knowledge Bank website; Call Center (1920) for Agriculture Advisory Service; e-SMS Service; Govi Mithuru project; and market price information systems [5].

Mauritius received financial support (Le Project) of 3 million EUR from European Union for the period of 2017–2022 to strengthen the climate-smart agriculture with the goal of achieving of UN Sustainable Development Goals on fighting climate change and its impacts, ending poverty and achieving food security. These kinds of funding and financial assistance are providing hope to vulnerable smallholders adapt and implementation of smart agriculture practices at village level to get substantial income.

Availability of Experts

The availability of technically trained experts and scientists in developed countries is another major challenge for island states. Expanding smart agriculture at an exponential rate in developed countries is attracting more technical experts with attractive remuneration. Unavailability of such technically efficient experts in small island states like Indonesia, Sri Lanka, Fiji, Mauritius, etc., is a foremost challenge and creates a negative impact on interested investors.

Therefore, continuously providing advanced skill-based training is required and recommended to farm owners and laborers related to the handling of sensors, software or hardware management, data interpretation maintenance of electrical components, or other farm implementations with proper safety in the field. Moreover, a mechanism should be developed for capacity building and technology transfer among scientists, young researchers, and academia on advanced technologies in smart agriculture. Effective outreach activities may be devised for more and more use of smart agriculture initiatives. A lack of high-quality, accessible agricultural extension services is another hurdle that limits the use of smart framings in many island states [8].

Data Availability, Reliability, and Insecurity

The success of climate-smart agriculture (CSA) depends on the availability of reliable data related to environmental parameters indispensable for crop production. In developing countries including small island states, the reliability and the availability of the data are in question [1] due to the availability of the limited number of metrological stations and data collection technicians. Data collection tools in the collection centers are manually operated and not fully digitalized considering the cost. Moreover, collected data are not stored in online platforms for distant access.

Power cuts and electricity supply interruptions are frequent in many small island nations. If power failure occurred, as many digital sensors are working on batteries, the recording of data will be interrupted; therefore, reliable data will not be available for forecasting. In some cases, if any data are interpreted wrongly due to power failure or system failure due to network interruption, irrigation or fertilizer will not be supplied on time or temperature will not be regulated if not monitored manually. Furthermore, unexpected climate change is another root cause that critically interferes with the forecasting based on the available data.

While integrating technologies into the agricultural sector such as mobile devices, remote sensing, big data, cloud, analytics, cyber security, and intelligent systems, security issues like compatibility, heterogeneity, constrained devices, processing, and protection of massive data will be some of the big challenges. In the majority of the developing countries, a licensed version of the software is not used; therefore, there are more chances of data stealing or distortion of information by internet hackers or cyber-attack [9].

Big Data Storage and Network Connectivity

In smart agriculture, IoT-mediated tools and applications need high-performance hardware and software facilities for quality data collection and storage, data acquisition, and interpretation of valuable data. In small island nations, the unavailability of extended internet connection to the farm areas and villages as well as the unavailability of fast network connections are hampering immediate data collection and fast processing when using AI-guided UAVs in fields. Moreover, the majority of the agricultural lands are located in rural areas. Electricity and internet facilities are not extended or not even available in those reigns. In such a situation, implementation of costly smart facilities would not be possible or even many interested farmers are hesitating to transform their farms into smart considering such hurdles [2].

Agricultural Policy

Smart agriculture is evolving as one mainstream opportunity to adapt and mitigate climate change and variability [10]. Therefore, national and regional level policies, programs, plans, and strategies on agriculture and/or the environment are being focused to transform the agricultural policies of the respective countries toward smart and precision farming. Strong and feasible policy on the digital transformation of agriculture is essential for small island nations, as agriculture plays a significant role in the economy as a contributor to GDP, employment, food, and income of people. But, the majority of the agricultural policy of the small island nations needed revision to incorporate modern technology substantially. For example, in Sri Lanka and Mauritius, it is much clear that policy is lagging behind in introducing 4IR technologies in agriculture in relation to the sector which faces a growing number of challenges and constraints that include low productivity, poor product quality, climate change, etc. [11].

According to the World Risk Report, Mauritius ranked as the 14th country with the highest disaster risks and ranked 7th on the list of countries most exposed to natural hazards. Therefore, Mauritius as a least developing country needs substantial foreign investment or financial assistance to expand the climate smart agriculture. Sri Lanka is another small island expecting foreign investment and assistance to strengthen smart agriculture due to frequent occurring of unexpected droughts and heavy floods. In Indonesia, it is reported that fertilizer subsidies by the government are resulting in over application of fertilizers in commercial agriculture, thus causing environmental damage [12]. Therefore, the Asian Development Bank (ADB) reported that fertilizer subsidy schemes provided to the farmers need to be reinvested into smart agricultural research, extension, and irrigation programs in order to promote the climate-smart agriculture to reduce the greenhouse gas emissions.

Strategies to Strengthen the Smart Agriculture

Small island nations are at risk due to the unforeseen extreme weather changes, erratic rainfall pattern, and expansion of soil salinity due to sea level increase. The following specific strategies are needed to strengthen the smart agriculture in these developing small island nations:

1. **Awareness creation**: Organizing regular frequent workshops on smart agriculture and digital farming, and its impact on farm productivity and profitability by the relevant government departments is important.
2. **Strengthen the finance support services**: Government should take initiatives to flow the substantial funds toward modernization of agriculture with the help of financial donors like ABD, World Bank (WB), Japanese International Cooperation agency (JICA), SADC, AU, UN, African Development Bank, FAO, UNESCO, UNDP, WHO, Afro-Asia Rural Development Organization, etc.

Respective governments should also arrange long-term or short-team loans or credit facilities to support farmers who are very much interested in investing money to expand the smart agriculture practices.

3. **Strengthen the island nations specific R&D**: Governments and all other stakeholders should encourage and investment in R&D on island nations specific smart agricultural technologies such as precision farming, Internet of Things (IoT), big data analytics, remote sensing and low-cost sensors, unmanned aerial vehicles (UAVs or drones), robotics.

4. **Appropriate policy reforms**: Governments of the respective island nations should adopt innovative appropriate policies that support opportunities in smart agriculture.

5. **Hiring facilities of smart tools**: Government should initiate renting facilities of smart tools like robots for fertilizer applications, UAV tractors for harvesting, etc., with the help of private partners; therefore, farmers need not to invest fully, and whenever needed, they can hire in low costs.

6. **Free database access**: Establishment of an interactive digital platform to allow farmers full access to climatic information, technology databases, expert systems, and DSS for web-based agro-advisory, skill development, machinery management, and financial assistance freely.

7. **Fast internet facilities**: A dedicated reliable, high-speed internet connectivity and communication network infrastructure should be made available, especially in rural areas as an enabler for development of smart agriculture.

8. **Extended power supply**: Uninterrupted power supply should be provided to remote farm lands for adopting smart farming technologies, for which renewable energy sources such as solar and wind energy should be extensively used.

9. **Human resource development**: A mechanism should be developed for capacity building and technology transfer among scientists, young researchers, and academia on advanced technologies on smart agriculture; therefore, shortage of technical personalities will be overcome.

10. **Land policy reforms**: Many farmers in the small islands do not have own land. To boost the smart agriculture, long-term lease facilities for agriculture lands or farmers should be encouraged to join cooperatives or other social institutions through policy interventions for applying smart farming technologies in agriculture.

Conclusion

The world is in the digital era. Modernization of agriculture is really important to minimize the land degradation and deterioration and greenhouse gas emission, safeguard the biodiversity, and increase the farm productivity and profitability. Smart agriculture is expanding rapidly to hop-up from the unforeseen climate change challenges and tackle current food security. Small island nations are most vulnerable to unpredictable climate change; therefore, there is an urgent need to transform the small island nation's agriculture toward smart. Initial cost to be incurred in implementation of smart agriculture is not affordable by majority of the farmers in the small island nations. There are specific challenges too to expand the smart agriculture in these small island nations. Careful analysis of problems and implementation of suitable strategies described in the chapter would be helpful to successfully transform the small island agriculture toward smart very quickly.

References

1. Rettore de Araujo Zanella, A., da Silva, E., Pessoa Albini, L.C.: Security challenges to smart agriculture: Current state, key issues, and future directions. Array **8**, 100048 (2020). https://doi.org/10.1016/j.array.2020.100048
2. Udutalapally, V., Mohanty, S.P., Pallagani, V., et al.: sCrop: a novel device for sustainable automatic disease prediction, crop selection, and irrigation in internet-of-agro-things for smart agriculture. IEEE Sens. J. **21**, 17525–17538 (2021). https://doi.org/10.1109/JSEN.2020.3032438
3. Autio, A., Johansson, T., Motaroki, L., et al.: Constraints for adopting climate-smart agricultural practices among smallholder farmers in Southeast Kenya. Agric. Syst. **194**, 103284 (2021). https://doi.org/10.1016/j.agsy.2021.103284
4. Rodriguez, J.M., Molnar, J.J., Fazio, R.A., et al.: Barriers to adoption of sustainable agriculture practices: change agent perspectives. Renew. Agric. Food Syst. **24**, 60–71 (2009). https://doi.org/10.1017/S1742170508002421
5. Narmilan, A., Niroash, G., Puvanitha, N.: Assessment of current status on smart farming technologies in Batticaloa District, Sri Lanka. Sri Lankan J. Technol. (SLJoT) **1**, 14–20 (2020). http://ir.lib.seu.ac.lk/handle/123456789/5171
6. Misra, N.N., Dixit, Y., Al-Mallahi, A., et al.: IoT, big data, and artificial intelligence in agriculture and food industry. IEEE Internet Things J. **9**, 6305–6324 (2022). https://doi.org/10.1109/JIOT.2020.2998584
7. Yogarajah, V., Weerasooriya, S.: Associated farm and farmer-specific factors for climate-smart agriculture adaptation in the dry zone: a case from Vavuniya, Sri Lanka. Sri Lankan J. Agric. Econ. **21** (2020). https://doi.org/10.4038/sjae.v21i1.4642
8. Mehta, C., Chandel, N., Rajwade, Y.: Smart farm mechanization for sustainable Indian Agriculture. Ama, Agric. Mechanization Asia, Africa Latin Am. **50**, 99–105 (2020)
9. Idoje, G., Dagiuklas, T., Iqbal, M.: Survey for smart farming technologies: challenges and issues. Comput. Electr. Eng. **92**, 107104 (2021). https://doi.org/10.1016/j.compeleceng.2021.107104
10. Partey, S.T., Zougmoré, R.B., Ouédraogo, M., et al.: Developing climate-smart agriculture to face climate variability in West Africa: challenges and lessons learnt. J. Clean. Prod. **187**, 285–295 (2018). https://doi.org/10.1016/j.jclepro.2018.03.199
11. Anonymous: Smart Digital Farming in Agriculture: Status and Prospects for Sri Lanka, 'Sri Lanka: State of the Economy 2019'. Sri Lanka, Institute of Policy Studies of Sri Lanka (IPS), pp 1–2 (2019). https://www.ips.lk/wp-content/uploads/2019/10/SMART-DIGITAL-FARMING-IN-AGRICULTURE-12.pdf
12. Savelli, A., Atieno, M., Giles, J., et al.: Climate-Smart Agriculture in Indonesia. CSA Country Profiles for Asia Series (2021). https://hdl.handle.net/10568/114898

Printed in the United States
by Baker & Taylor Publisher Services